OUTSTANDING DISSERTATIONS IN LINGUISTICS

Edited by

Laurence Horn

Yale University

A ROUTLEDGE SERIES

Outstanding Dissertations in Linguistics

Laurence Horn, *General Editor*

The Effects of Duration and Sonority on Contour Tone Distribution
A Typological Survey and Formal Analysis
Jie Zhang

Existential Faithfulness
A Study of Reduplicative TETU, Feature Movement, and Dissimilation
Caro Struijke

Pronouns and Word Order in Old English
With Particular Reference to the Indefinite Pronoun Man
Linda van Bergen

Ellipsis and wa-marking in Japanese Conversation
John Fry

Working Memory in Sentence Comprehension
Processing Hindi Center Embeddings
Shravan Vasishth

Input-based Phonological Acquisition
Tania S. Zamuner

Vietnamese Tone
A New Analysis
Andrea Hoa Pham

Origins of Predicates
Evidence from Plains Cree
Tomio Hirose

Causes and Consequences of Word Structure
Jennifer Hay

The Phonetics and Phonology of Gutturals
A Case Study from Ju'hoansi
Amanda Miller-Ockhuizen

True to Form
Rising and Falling Declaratives as Questions in English
Christine Gunlogson

Phonological Augmentation in Prominent Positions
Jennifer Smith

Category Neutrality
A Type-Logical Investigation
Neal Whitman

The Ups and Downs of Child Language
Experimental Studies on Children's Knowledge of Entailment Relations
Andrea Gualmini

Markedness and Faithfulness in Vowel Systems
Viola Miglio

The Syntax-Information Structure Interface
Evidence from Spanish and English
Eugenia Casielles-Suarez

Lenition and Contrast
The Functional Consequences of Certain Phonetically Conditioned Sound Changes
Naomi Gurevich

Lenition and Contrast

The Functional Consequences of Certain Phonetically Conditioned Sound Changes

Naomi Gurevich

Routledge
Taylor & Francis Group
NEW YORK AND LONDON

Published in 2004 by
Routledge
Taylor & Francis Group
711 Third Avenue, New York, NY 10017
2 Park Square, Milton Park, Abingdon, Oxfordshire OX14 4RN

Routledge is an imprint of the Taylor & Francis Group, an informa business

Transferred to Digital Printing 2005

First issued in paperback 2013

10 9 8 7 6 5 4 3 2 1

Catalog record is available from the Library of Congress

ISBN 13: 978-0-415-97099-0 (hbk)
ISBN 13: 978-0-415-86514-2 (pbk)

This book is dedicated to Tim, Hamish, the Office and a coaster.

Acknowledgments

I couldn't have done this without my advisor and friend, Daniel Silverman.

Table of Contents

Lenition and Contrast

Outstanding Dissertations in Linguistics

CHAPTER 1

Introduction

In this study I investigate 230 phonetically conditioned sound changes, mostly lenitions, and find that there is an overwhelming tendency for these phenomena to avoid neutralization. This suggests that such processes do not operate independently of functional considerations.

For the sake of clarity I start by explaining my use of *phonetically conditioned sound changes*, *functional considerations*, and *neutralization*: PHONETICALLY CONDITIONED SOUND CHANGES are changes affected by the physical properties of articulators and their acoustic consequences. That is, sound changes that can be attributed to the shape and/or the movement of the vocal organs are considered phonetically conditioned. Since human vocal organs and the influence they bear on acoustics are the same regardless of one's native language, sound changes that are attributed to these factors are rather widespread cross-linguistically. Two common examples of such changes are inter-sonorant voicing in Ao (§3.3), as illustrated in data (1.1), and final devoicing in Bulgarian (§3.17), as illustrated in (1.2).

(1.1) Voicing of stops in Chungli dialect of Ao (Gurubaswe-Gowda 1975: 11):

a.	apu˩˥	[abu˩˥]	'blow'
b.	a˥˩tep	[ʌ˥˩dep]	'patch'
c.	loŋki	[lɔŋgi]	'hole (made by insect)'

(1.2) Final devoicing of obstruents in Standard Bulgarian (Scatton 1984: 73):

a.	grad-ove	'cities'
b.	grat	'city'

A simplified description of the physical influences on inter-sonorant voicing is that the vocal cords continue to vibrate after the first sonorant, through the obstruent, and into the second sonorant. In word-final position, however, the aerodynamic conditions are not conducive to voicing (Westbury and Keating 1986).

FUNCTIONAL CONSIDERATIONS refer to meaning and how it is affected by sound substitutions, which may either obliterate meaning distinctions or maintain them. The term *functional* derives from Trubetzkoy's (1939) definition of FUNCTIONAL LOAD: the number of elements in the lexicon that depend on a particular opposition in order to be distinguished from others. Thus, the system of contrasts in a language and the meaning distinctions of lexical items that depend on these contrasts are considered matters of *function*: these distinctions affect communication, which is the primary *function* of language.

Sound substitutions can affect communication by either maintaining distinctions between PHONEMES (sounds that serve a contrastive function in a given language) or eliminating them. Sound substitutions where a phoneme is replaced with a sound that does not represent any other phonemes in the language do not obliterate meaning distinctions. Such substitutions are never neutralizing because they do not interfere with distinctions between phonemes. An example of such a process is the non-initial partial voicing of stops in Southern Italian (§3.48), as illustrated in (1.3). The partially voiced b̥ d̥ g̊ surface as non-initial ALLOPHONES (phonetic variants of phonemes) of p t k, and since they are not phonemic in the language, nor do they serve as allophones of any other phonemes, no contrast between phonemes is affected and neutralization is avoided.

(1.3) Examples of partial voicing in Southern Italian (Oftedal 1985: 36):

a.	['parte]	'part'	[di 'b̥arte]	'of a part'
b.	['terra]	'land'	[la 'd̥erra]	'the land'
c.	['karne]	'meat'	[di 'g̊arne]	'of meat'

Sound substitutions where a phoneme is replaced by a sound that is either phonemic in the language, or an allophone of another phoneme, may neutralize contrast between phonemes. That is, if more than one phoneme is represented by the same sound, there is an obliteration of the phonetic distinction between the phonemes in question (i.e., PHONETIC NEUTRALIZATION), where both are pronounced and sound the same. An example of a process that results in phonetic neutralization is the word-final devoicing of b in Yakut (§3.149):

(1.4) Final devoicing of b in Yakut (Krueger 1962: 56):

a.	saba	'his thread'	sap	'thread'
b.	tabar	'he hits'	tap-	'to hit'

Since both b and p are phonemic in Yakut, the word-final substitution of b with p results in the phonetic neutralization between these two phonemes. However, phonetic neutralization in itself does not necessarily lead to the obliteration of meaning distinction between words (i.e., the obliteration of LEXICAL DISTINCTION). In Yakut p is a marginal phoneme which is rare and occurs mostly in non-native words, and furtheremore, is never attested word-finally. Hence, all word-final occurrences of p can only be manifestations of the b phoneme, and the p/b opposition is maintained because there is never a case where the phonemic association of a surface p could be misinterpreted. In other words, all non-final occurrences of p represent the p phoneme, and all final occurrences represent the b phoneme.

If phonetic neutralization between sounds which represent different phonemes occurs in the same context, the consequence for communication is potentially crucial: such neutralization leads to the possibility that a sound could represent more than one phoneme in a language, with no way to determine which. This obliterates the distinction between the phonemes in question. I term this PHONOLOGICAL NEUTRALIZATION. Such neutralization can potentially lead to the obliteration of lexical distinction, which could result in the functional consequence of hindering communication because it could lead to confusion between lexical items that sound the same (i.e., it could result in HOMOPHONY).

An example of a process that leads to phonological neutralization is the spirantization of t in Nez Perce (§3.88):

(1.5) t → s / _n, _w in Nez Perce (Aoki 1970: 39):

a.	[ˈjuʔt]	'poor', 'pitiful'	[ˈjuʔsne]	'poor' (object case)
b.	[ʔojˈlaːqt]	'six'	[ʔojˈlaːqswa]	'six' (of men)

The substitution of s for t in certain contexts obliterates the phonetic distinction between these two sounds, both of which are phonemic, because both surface as s. Since in Nez Perce phonemic s could also occur in _n and _w environments, there is no way to determine which phoneme is represented by a surface s in clusters such as sn and sw. This results in phonological neutralization because the loss of phonetic contrast between t and s leads to a possible obliteration of meaning distinction between words.

To summarize the distinctions I make among different kinds of neutralization: if a sound represents only one phoneme in a language there is *no neutralization*. If a sound represents more than one phoneme in a language there is *phonetic neutralization*. Phonetic neutralization that results in confusion between phonemes is *phonological neutralization*. For any of these to be truly consequential for communication they must result in *homophony*. See §2.1.3.1

for a discussion of why phonological neutralization in itself is only potentially consequential to lexical distinction.

In this study I show that certain phonetically conditioned sound substitutions, especially lenition processes, are far more likely to maintain meaning distinctions than they are to obliterate them. I conclude that processes that can be attributed to biological, rather than phonological, factors are consequently constrained by the system of contrasts in a language. In other words, processes that can be explained by what is common among humans (i.e., the properties of the vocal organs) are constrained by functional considerations, which are language-specific.

1.1. Study Overview

The role that contrast plays, if any, in phonetic processes can only be assessed by examining the meaning-maintenance status of a large number of such processes from a wide variety of languages. While the motivation behind such processes can only be hypothesized, their functional consequence can, in most cases, be assessed. For this purpose I investigate 230 phonetically conditioned processes, mostly lenitions, in 153 languages and examine the ratio of phonologically neutralizing to non-neutralizing processes. Indeed, 92% avoid phonological neutralization. I conclude that the tendency for phonetically conditioned processes such as lenition to maintain contrast by avoiding neutralization is so overwhelming that the dependence between such processes and contrast is unambiguous.

1.1.1. Collection of Data

My typology is based on a database of lenition processes compiled by Lavoie (1996), augmented by Kirchner (1998), and presented in the appendix of Kirchner's dissertation as examples of processes that fit his definition of Lenition. LENITION (Chomsky and Halle 1968, Vennemann 1969, Hock 1976; 1991, Kirchner 1998, Lavoie 1996) or WEAKENING, is a cover term for a variety of both synchronic and diachronic changes that for consonants are often "some reduction in *constriction degree or duration*" (Kirchner 1998: 1). There are no clear guidelines in the field for what exactly constitutes lenition but certain generalizations do emerge as vital, such as the fact that an element's strength is context-dependent (i.e., the same sound may be considered more or less weak depending on its environment). Agreement on the motivation behind lenition processes is also lacking but it is widely assumed that lenition is driven by phonetic considerations, specifically, ease of production and perception (e.g., Steriade 2000, Kirchner 1998, Jun 1995, Flemming 1995).

Compiling a list of phonetically-conditioned processes for the purpose of the present study could conceivably taint the statistical analysis, exposing the study to the criticism that possible cases of neutralization could have been purposely excluded in order to raise the ratio of non-neutralizing processes and better the statistics. For this reason it is preferable to use a pre-compiled corpus, where contrast matters do not serve as factors for either inclusion or exclusion in the collection of processes. Since neither Lavoie nor Kirchner emphasizes contrast as relevant to their research, using their inventory of lenition processes does not compromise the legitimacy of the study's results.

In using Kirchner's compilation I systematically investigate the functional properties of every process in the same source cited by Lavoie or Kirchner, excluding processes only when necessary and based on strict criteria that are, again, unbiased with respect to the focus of the study. That is, no process is ever excluded based on contrast matters, but only based on the ability to judge whether it is neutralizing or not. This determination depends wholly on the availability of sources and similar objective concerns (which are detailed in §2.1).

I investigate the processes documented in Kirchner's work, but I make no attempt to explore the history of these phenomena. For example, in Gujarati (§3.44) the retroflex stops are flapped, which does not neutralize any contrasts in the language. However, the retroflex stops in Gujarati (and other Indo Aryan languages) resulted from a diachronic sound merger between voiced and voiceless retroflex stops (Hans Henrich Hock, p.c.), which suggests that the sounds undergoing this non-neutralizing process of flapping may have taken part in a neutralizing, at least phonetically, process earlier in the history of the language.

On a similar note, the spirantization of b d g in Iberian languages (e.g., in Catalan (§3.20) and Spanish (§3.120, §3.121)) is classified as non-neutralizing, but its effect on the bilabial stop may have contributed to the obliteration of the distinction between sounds that represented the old Latin voicing contrast (p/b).[1] However, although it is not clear whether the merger between Old Spanish b and β resulted in phonological neutralization, it is certain that the processes included in the present corpus, i.e., the spirantization of stops in Catalan and Spanish, do not obliterate distinctions between any current phonemes. Hence, although the history of the sounds undergoing the processes in my corpus is interesting, it is beyond the scope of the present study.

---

[1] Old Spanish had a phonemic contrast between intervocalic β (< Latin b) and b (< Latin p), which was apparently neutralized when b → β (Penny 2002).

In addition to this, several processes in the corpus are related and could potentially be viewed as parts of the same phenomenon. For example, the various processes in individual languages that belong to the more general phenomenon of flapping the retroflex stops in Indo Aryan languages, or the spirantization of stops in Romance languages, as well as the intervocalic flapping of alveolars in various dialects of English, could all be analyzed as instances of the same general processes. However, conflating related processes is only possible in the case of the more commonly studied languages whose histories are well documented, but not with the lesser studied ones. Such a practice would treat more commonly studied languages differently from others, and therefore would bias the statistical analyses in the present study. For this reason I follow Kirchner in listing each instance of a process in each language as a separate entry in the corpus.

Finally, although new information is sometimes added when it is essentially related to a process, no new processes are added to the inventory in order to avoid tainting the statistical analysis of a pre-existing compilation. In this way the results are kept as unbiased as possible with respect to the focus of the present research: functional considerations and how they constrain phonetically conditioned processes.

1.2. State of the Field

The functional considerations of phonetically conditioned changes are rarely explored in literature, although some relevant discussion exists. In this section I briefly discuss work on the relationship between meaning distinctions and phonetically conditioned sound substitutions (Jacobs and Wetzels 1988, Jacobs 1994, Silverman 2000, Hualde 2000, Hyman 1999), research on phonetically conditioned processes removed from functional considerations (Kirchner 1998), and treatment of lenition processes within markedness-based theories (based on work by Trubetzkoy 1939, Chomsky and Halle 1968, and others).

1.2.1. Jacobs and Wetzels (1988), Jacobs (1994)

In their analysis of diachronic lenition in French Jacobs and Wetzels (1988) stress the non-neutralizing aspect of this phenomenon. Historically, French intervocalic obstruents were voiced, then spirantized, before eventual gliding and deletion (1.6).

(1.6) French Lenition (Jacobs 1994: 2):

a.	fratrem	>	*[fradre]	>	[fraðre]	>	frère	'brother'

b. aprilem > *[abrile] > [aβrile] > avril 'April'

The fact that the spirantization of voiced stops did not result in contrast obliteration, because no voiced spirants existed in the inventory of Latin sounds, is seen as indicative of lenition processes in general. Jacobs and Wetzels (1988) conclude that the tendency of lenition processes to avoid neutralization must therefore be accounted for in the formalizations of such phenomena.

For this purpose lenition is described in terms of underspecification and default fill-in rules. A simplified explanation of how this works is, for example in (1.6a) where t > *d > ð, the feature [continuant] is seen as underlyingly unspecified in the case of d because, at that stage in the language, no distinction between continuants and non-continuants existed for voiced obstruents. Thus, the change from d to ð is formalized as a move from [0cont] to [+cont], and as such, it is a feature-filling rule (i.e., filling in a value from the previously unspecified [0] to the specified [+]). This formalism captures the non-neutralizing tendency of French lenition because feature-filling rules add, rather than eliminate information and therefore can not obliterate contrast.

One of the problems with the underspecification formalization of lenition, which Jacobs (1994) addresses, is its inability to straightforwardly account for synchronic allophonic lenition processes where voicing and spirantization take place simultaneously. This, and other, problems with the underspecification account are discussed and solved by turning to an Optimality Theoretic (OT) (Prince and Smolensky 1993) account of French lenition.

Jacobs' intuition, albeit based on a single phenomenon in Romance, that lenition processes are mostly non-neutralizing is borne out in the present study. The motivation behind such processes, and their mostly non-neutralizing consequences, are not attributed to a single formalization however. Where Jacobs strives to include the non-neutralizing tendency of lenition processes as part of the phonological rules implemented by speakers, I prefer to regard the same tendency as a gradual, diachronic, force that interacts with processes attributed to phonetic causes.

1.2.2. Silverman (2000)

Silverman (2000a) proposes that while certain allophones may result from phonetically-induced changes, the systemic responses to such changes are not always phonetically grounded, but instead are often a matter of contrast maintenance. For example, consider intervocalic voicing and intervocalic spirantization in Corsican (1.7, 1.8).

(1.7) Voicing in Corsican (Dinnsen & Eckman 1977: 6):

a.	[peðe]	'foot'	[u beðe]	'the foot'
b.	[tengu]	'I have'	[u dengu]	'I have it'
c.	[kaza]	'house'	[a gaza]	'the house'

(1.8) Spirantization in Corsican (Dinnsen & Eckman 1977: 6):

a.	[bokka]	'mouth'	[a βokka]	'the mouth'
b.	[dente]	'tooth'	[u ðente]	'the tooth'
c.	[gola]	'throat'	[di ɣola]	'of throat'

Neutralization between voiced p t k and existing b d g is avoided because in the same context where p t k → b d g, phonemic b d g → β ð ɣ.

In the Corsican example the sounds b d g belong to two sets of phonemes: they are phonemic in themselves (1.8), and also represent a phonetic variation of the p t k phonemes (1.7). Neutralization between the two sets of phonemes represented by b d g is avoided because in the same context where p t k surface as voiced, b d g surface as spirants (β ð ɣ). This is an example of PHONEMIC OVERLAP, an intersection of phonemes where "a given sound [...] may belong to two or more different phonemes in the same dialect" (Bloch 1941: 93). In contrast to cases of neutralization, which necessarily involve an overlapping between phonemes and their phonetic variants, phonemic overlap does not lead to the obliteration of meaning distinctions because in the same context where one phoneme sounds like another, the second phoneme is realized as a different phonetic variant. Thus, in the case of Corsican, all intervocalic b d g must represent the p t k phonemes because in this context phonemic b d g surface as β ð ɣ.

The results of the present study are consistent with Silverman's prediction that phonetically-induced changes may result in additional changes motivated by contrast-maintenance, examples of which can be found in the discussion of phonetic, but not phonological, neutralization in the corpus, and how meaning distinctions are maintained in each case (§2.3.2.3).

1.2.3. Hualde (2000)

Hualde (2000) writes that an initial phonetically motivated step in lenition chains and vowel shifts often leads to systemic or phonologically motivated changes:

> In such chains, although the initial step may be phonetically motivated, the rest of the changes would rather appear to be systemically or phonologically motivated, with the result that merger between phonemes is avoided and, in

> some sense, the 'same' previously existent contrast is preserved after the change. (p.99)

Hualde contends that, although some chain-like phenomena may have consisted of unrelated processes that occurred at different times, "chain-like systemic sound changes are a fact of linguistic evolution" (p.99).

Hualde goes on to argue that the extensive reconstruction of the accentual system in Markina Basque is an example of such a system-driven change. In Markina Basque there was a phonetically-motivated accent shift from final to penultimate position (1.9a), and a subsequent shift of lexical accent to antepenultimate (1.9b). Note that the second shift helps preserve the contrast between singular and plural forms of some lexical items in the language.

(1.9) Accent shifts in Markina Basque (Hualde 2000):

a.	lagu'ne	>	la'gune	'the friend'
b.	la'gunek	>	'lagunek	'the friends'

Hualde stresses that the way in which contrast is preserved in Markina Basque is indicative of the way in which stability is restored to a linguistic system when it is threatened by phonetic sound change, but is by no means an inevitable, or even predictable, consequence of such threat. The author proposes that the second accent shift (1.9b) was triggered by the reestablishment of the generalization "the plural is accented one syllable earlier than the singular" following the first shift (1.9a). While Hualde does not view this shift as triggered by the implicit need to avoid neutralization, the preservation of a difference that keeps two forms distinct (in this case, the difference in which syllable is accented depending on whether the form is singular or plural) is paramount to contrast maintenance. Similar systemic changes that help preserve a contrast in danger of obliteration are discussed in the present study. In Chapter 2 several such merger-avoidance strategies are presented.

1.2.4. Hyman (1999)

Hyman (1999) expresses the view that lenition and fortition processes operate independently of contrast: "…'Contrast' seems to play little or no role in such processes. The same Bantu fortition or lenition such as the l/nd alternation, can either be neutralizing in one language, e.g. Yao, vs. allophonic in another, e.g. Yaka."

Hyman's observation is limited to one particular process and its manifestations in various Bantu languages, and should probably not be interpreted as an argument against the general role of contrast in phonetically-conditioned changes across languages. However, it does seem to echo the general intuition in the field, an intuition that is rarely addressed in literature, that phonetic changes operate independently of contrast considerations. The results of the present study, based on 230 processes, suggest that contrast must in fact play a role in phonetically conditioned processes because so many of them avoid neutralization.

1.2.5. Kirchner (1998)

Kirchner (1998) proposes an optimality-theoretic based "effort minimization theory" which ranks effort considerations with respect to other constraints, creating a counter force between effort minimization constraints (LAZY) and faithfulness and fortition constraints. Although he does not specifically acknowledge meaning maintenance considerations as relevant to his theory, they do nevertheless play a role: contrast is inherently related to faithfulness constraints. Within OT faithfulness constraints prohibit distinction between an underlying value (i.e., the phonemic representation of a word) and its surface realization. Every difference between a candidate and the phonemic representation of a word is considered one violation of the faithfulness constraint, which, in theory, keeps candidates that minimize effort but maximize violations of faithfulness from winning.

Nevertheless, Kirchner specifically argues against formalizing lenition rules in terms of contrast because this view fails to capture the naturalness of the more common lenition processes such as spirantization or voicing. Formalization difficulties aside, the present study shows that contrast does in fact play an integral part in such processes. In viewing lenitions as phonetically induced but phonologically constrained I am able to account both for the naturalness of certain processes (attributed to their phonetic motivation) and for the role that contrast plays (one of constraining the spread and consequences of phonetic changes).

1.2.6. Markedness-Based Theories

MARKEDNESS terminology entered the field of phonology with Trubetzkoy's (1939) classification of neutralization types, and his empirically-based observation that context dependent neutralization usually results in the unmarked member of an opposition. He goes on to implement markedness values as part of the hypothesis that the frequency of sounds depends on their degree of articulation complexity (Zipf 1935). Thus, in using markedness values as linguistic representations of articulation complexity, which Trubetzkoy considered extra-linguistic and therefore inappropriate as part of phonological theory, he initiates the association between markedness and some degree of "naturalness."

Subsequently, within generative theories markedness becomes associated with linguistic universals and comes to embody a measure of the naturalness of linguistic elements, where the less marked, the more natural (Chomsky and Halle 1968, Guitart 1976, McMahon 1994, Rice 1999, among others). At the heart of this abstract measure of naturalness remains Trubetzkoy's original observation that the result of neutralization is usually unmarked, which leads to the conclusion that neutralization is a move towards a more natural state.

Lenition, especially in the generative tradition, is viewed as a change towards a more natural, easier to produce, state. In fact, lenition is most commonly defined as a reduction in the degree of articulatory complexity (e.g., Steriade 2000, Kirchner 1998, Jun 1995, Flemming 1995). Just like neutralization, therefore, lenition is considered within markedness-based theories as a move towards the unmarked because it is defined as a more natural state, which leads to the conclusion that lenition and neutralization are somehow related. This conclusion contradicts my findings that lenition processes avoid neutralization 92% of the time.

1.2.7. Ongoing Issues

My primary contribution lies in establishing a relationship between phonetically conditioned sound changes and functional considerations. Contrast is an integral part of phonology, but it is rarely emphasized in discussions of phonetically conditioned sound changes. The results of this study show that the relationship between phonetic changes and matters of contrast is unambiguous, and therefore, its examination fills a gap in existing literature.

In addition to showing that phonetically conditioned sound changes are indeed constrained by contrast, the current study serves to illustrate previously unidentified trends in the functional aspects of phonetic changes. The most striking of these trends is that, statistically, lenition phenomena that result in phonetic neutralization rarely result in phonological neutralization. In phonology literature phonetic neutralization is often synonymous with phonological neutralization, where if the phonetic distinction between phonemes is obliterated the default assumption is that distinction between words will also be lost. The present study, however, shows that phonological neutralization is not necessarily the outcome of phonetic neutralization.

This trend is especially surprising in the case of LOSS or DELETION of a phoneme, i.e., where a phoneme is realized as Ø in a certain environment. For example, in Bulgarian (§3.17) nasals are lost between a vowel and a fricative (1.10).

(1.10) Loss of nasal in Bulgarian (Scatton 1984: 57):

onzi	[õzi]	'that'

In Bulgarian the phonetic distinction between the existence of n and its absence (Ø) is lost where n is deleted. However, the contrast between the existence of the phoneme n and its absence is maintained by shifting to the previous vowel, which is nasalized, and phonological neutralization is avoided (i.e., if ozi is a word in Bulgarian, it remains distinct from õzi). Thus, by distinguishing between *phonetic* and *phonological* neutralization, and showing that the first does not necessarily result in the second, I uncover a previously unexplored trend in the relationship between phonetics and phonology.

The current study is based on empirical observations in data rather than theory-internal motivations. In the same spirit the terminology, especially terminology that implies in its definition affinity with certain phonological theories, is regularly defined in the text and summarized in the Glossary (Appendix B) so that my discussion and intentions are made explicit. Thus, the validity of the statistical trends analyzed in the current study should remain uncompromised regardless of the changing theories in the discipline of phonology.

1.3. Structure of the Book

In Chapter 2 the study is discussed in detail: methods used to collect data are explained, results are illustrated, and the various trends observed in the statistics are discussed. Chapter 3 is the analysis of every process included in the study, indicating the reasons for classifying each process as either neutralizing or not. In Chapter 4 the implications of the present study, and its findings, are presented, and questions requiring further research are identified.

The first appendix is a table that summarizes the main results of the study, listing each process and its functional classification. Finally, throughout the book definitions of linguistic terms are included. The first use of each defined term is printed in SMALL CAPS, and all such terms are summarized in the Glossary which is presented in Appendix B.

CHAPTER 2

Investigation of Phonetically Conditioned Sound Changes

2.1. Methods

2.1.1. Grammar Sources

In order to examine the relationship between contrast and phonetic changes I research 230 such processes and determine whether or not they bring about phonological neutralization. Each process is analyzed using the same source cited by Kirchner (1998). Using original sources is important because, although they describe the same phenomena, grammars of the same language do not always agree on details, especially when it comes to the contrastive properties of sounds. For example, consider the analysis of Bontoc (§3.16) where Lavoie's (1996) original source (Reid 1971) lists p t k and b d g as 6 separate phonemes, while another source on the language (Sadenadel 1909) describes the voiced and voiceless variations of each stop (p/b t/d k/g) as members of the same phonemes. Since the inventoried process in Bontoc (b d g → f ts k^h / syllable-initially) originates in Reid's work, and in the absence of additional data on the language, his description carries more weight.

For these reasons it is important to look up each process in the very source in which it is cited so that (1) the existence, and description, of each process can be verified; (2) the functional consequences of each process can be researched using the contrastive information provided by the same source that lists it, thus avoiding inconsistencies stemming from varying interpretations of language data; and (3) a consistent strategy for the use of grammar sources is followed throughout the study, and is not dependent on my ability to judge the validity of grammar sources.

On occasion Lavoie and Kirchner use secondary sources to report on a language, but whenever possible I consult the primary source. Additional sources on some grammars are consulted on a need-to basis. For example, in the case of Faroese (§3.33) the source Kirchner cites for the intervocalic and pre-liquid voicing of stops (p t c k → b d ɟ g / V_V, _liquids), Lockwood 1964, provides very little information on the contrastive property of the relevant sounds. For this reason 3 additional sources are consulted on Faroese: Strodach 1976, Barnes 2001, and Staksberg p.c.

2.1.2. Explanation of Changes to Data

Kirchner (1998) writes that his collection includes 272 languages, however, only 193 languages are listed in his appendix. Of these, 16 are omitted due to unavailable or unusable sources. This includes sources that are in a language foreign to me as well as a number of unpublished sources, and in very rare cases sources that were simply not available through the Library Gateway at the University of Illinois at Urbana-Champaign which includes access to off-campus catalogs such as ILLINET Online, a resource sharing system for the 44 ILCSO (Illinois Library Computer Systems Organization) libraries, as well as interlibrary loan available through IRRC (Information Resource Retrieval Center) that searches the catalogs of the Big Ten institutions, among others.

An additional 19 languages are omitted for a variety of reasons that span from a lack of relevant information (that is, an inability to determine the functional consequences of the processes in a language) to not finding the processes in question in the cited sources for the language. Finally, 5 languages are omitted because they are repeats (that is, cases where the same language, and sometimes the same process in the same language, are listed as separate entries in Kirchner's inventory). This results in 153 languages that are used in the present study.

230 processes in these 153 languages are investigated. These processes are listed as closely as possible to the manner in which they are presented in the cited sources, which in some cases differs from Kirchner's descriptions. In addition to rewriting some of the processes, I omit those processes that I was not able to find in the source they are attributed to.

Several diachronic processes are also omitted because their functional consequences are, in some cases, difficult to classify. As far as statistical analyses of trends in the corpus are concerned, no distinction is made between synchronic and diachronic processes. However, diachronic processes are sometimes eliminated, not due to their diachronic nature but because no information regarding the system of contrasts of the proto-language in question is available. Without such information it is usually not possible to determine whether a particular process neutralized any contrasts in the language or not.

For example, in Proto-Ainu (§3.97) there is a diachronic processes that is listed in Kirchner as 'delete h, probably V_V', and is described in the grammar source as *h > h/Ø (Vovin 1993). No contrast information about the language at the stage where *h was part of its inventory is available, or what conditions determined whether *h > h or *h > Ø. It would appear that, since the phonetic distinction between the existence of h and its absence (Ø) is neutralized, *h > Ø should automatically be classified as neutralizing. However, most other cases of deleted h's in the present corpus have been shown to avoid phonological neutralization (Table 1 shows that 6 of the 10 such cases in the corpus are classified as not-neutralizing).

Table 1: Functional Consequences of h Deletion in the Corpus

Not Neutralizing		*Neutralizing*	
Mataco (Noctenes)	? h - Ø	Pawnee 1	h - Ø
Nepali	h - Ø	Sanuma 3	h - Ø
Newari	h - Ø	Tharu (Chitwan)	h - Ø
Nez Perce	h - Ø	Tojolabal	h w j - Ø
Shoshone (Tümpisa)	? h - Ø		
Southern Tati (Chali)	h - Ø		

In the Chali dialect of Southern Tati (§3.116), for example, the loss of h is accompanied by compensatory lengthening of the preceding syllable, which suggests that the distinction between the existence of h and its absence remains intact (that is, the contrast shifts from segmental h to the length of the preceding syllable; see §2.3.2.3 for a detailed discussion of such phenomena). Hence, it is not possible to determine whether the same type of process in Proto-Ainu resulted in the neutralization of any contrasts, and therefore, this process and several other diachronic processes under similar circumstances are omitted from the corpus.

Finally, in order to maintain the integrity of the statistical results which depend on how many processes are neutralizing and how many are not, I take care to group parts of the same process as one item. For example, Kirchner lists the lenitions in Yindjibarndi as 9 separate processes (2.1).

(2.1) 9 lenition processes listed for Yindjibarndi (Kirchner 1998: 314):

a. p → w / V_V

b. k → w / u_

c. k → Ø / V_V

d. t̪ → ð̪ / V_V
e. tʲ → j / V_V
f. glides → Ø / V_V
g. t ʈ - ɾ
h. r → j / a_i
i. r → Ø / V_V

Of these I omit those processes that could not be found in Kirchner's source for Yindjibarndi grammar (Wordick 1982) and condense the rest into relevant groups, listing them as 2 processes in the present study (2.2).

(2.2) 2 processes listed for Yindjibarndi (§3.151):
a. p t̪ ɻ k → w j j w / V_V
b. ʈ → ɽ / V_V

2.1.3. Language Analyses

The analysis for each of the 153 languages is presented in Chapter 3, and is organized as follows:

2.1.3.1. Summary of Essential Information

At the start of the section for each language essential information is summarized into headings, as illustrated in Table 2.

Table 2: Essential Language Information

Language:	*Dialect*:
Phylum:	
Inventory:	*Source*:
Add. Source:	
Lenition(s):	
Type:	
Classification:	
Trends:	
Difference:	

The language name and the relevant dialect, if one is specified, are listed, as well as the language phylum in full. There is no one standard in the field of linguistics to draw on for lineage information on languages, and researchers use a variety of classification techniques. In order to provide consistent information, especially with respect to the relationship among the languages, I rely on a singular source of information on language phyla. SIL International's online Ethnologue is not only easily available, but is also incredibly broad in the number of languages it includes and the details it provides about each language's family. Therefore, all phyla information in the current study is cited from this source.

After the phylum of each language its origin from either Lavoie's (1996) or Kirchner's (1998) inventory is specified as well as the grammar source from which it is cited and a list of additional sources if any are used. Next, the processes that make up the database for the corpus are listed along with a general "type" for each process (i.e., whether it is spirantization, voicing, loss, gliding, etc.) and, of course, the functional classification of each process as either neutralizing or not. These classifications are divided into 4 types:

1. *Not Neutralization*: processes that do not result in phonological neutralization of any contrasts. E.g., in some dialects of Assamese (§3.9) word final aspirates p^h t^h $b^ɦ$ tend to spirantize to ɸ θ β. This does not lead to phonological neutralization because the spirants do not represent any other phonemes in the language other than the aspirates, so no distinction between phonemes is obliterated.
2. *Limited Neutralization*: processes that result in the phonological neutralization of a contrast, but its effect is so minor that it could not significantly affect communication. E.g., in Nez Perce (§3.88) post-consonantal h is deleted. This does neutralize the distinction between Ch and CØ, but its consequence for communication is minimal because it applies in an extremely limited environment: although clusters of up to 4 consonants are allowed in the language, the only context in which h can follow a consonant is as the 2nd member of a CC cluster when the first member is one of only 4 consonants (s, x, w and $q^ʔ$). When only a handful of lexical items are affected by a phonetic neutralization, meaning distinctions are rarely obliterated.
3. *Incomplete Neutralization*: processes that result in phonological neutralization which is limited in some respect. E.g., in Nepali (§3.86) b^h d^h g^h optionally de-aspirate to b d g everywhere except word-initially. This obliterates the distinction between the aspirated and plain voiced stops, all of which are phonemic, but is considered incomplete because of its interaction with other processes that serve to preserve contrast in various contexts: intervocalically for the bilabial and alveolar stops because in this context phonemic b d are laxed, so deaspirated b^h d^h should remain distinct, and in the vicinity of nasals

for all three because in this context phonemic b d g are nasalized, which should keep them distinct from deaspirated b^h d^h g^h.

4. *Neutralization*: processes that result in complete phonological neutralization. E.g., in Lama (§3.63) word-final p glides to w. Since both p and w are phonemic in the language, this obliterates the distinction between them because there is no way to tell which phoneme is represented by a word-final w.

The first two of these classifications are considered basically not neutralizing and add to the count of non-neutralizing processes in the corpus (207 not, and 5 limited, neutralizations total 212 processes) while the second two types add to the count of neutralizing processes (2 incomplete, and 16 neutralizing total 18 processes). The use of 4, rather than 2, classification types bears little theoretic significance, and is implemented merely for the sake of descriptive accuracy.

The classification of processes as either neutralizing or not is not equally simple in all cases. Avoidance of neutralization is often straightforward and can usually be established: if a phoneme is substituted with an allophone that does not represent any other phonemes in the language, no obliteration of contrast between phonemes is possible. Phonological neutralization, however, is often difficult to determine conclusively. In fact, none of the cases classified as neutralizing in the corpus can be conclusively established as obliterating lexical distinctions to the point of hindering communication.

Processes that result in phonological neutralization can potentially lead to homophony. However, this potential may never be realized. In order to determine whether the observed phonological neutralization actually leads to homophony one must have access to very detailed information about the lexicon and the grammar in question, information that is well beyond what most sources provide. Despite phonological neutralization, the obliteration of lexical distinction may be avoided in several ways:

1. *Absence of minimal pairs*: MINIMAL PAIRS (pairs of words that are distinguished only by a difference in one phoneme; e.g., Hebrew ʔani 'I' and ʕani 'poor' are minimal pairs distinguished exclusively by the ʔ/ʕ opposition) are the only real candidates for homophony if the distinction in question is neutralized. It is extremely rare to encounter a grammar source for a language that will list all possible minimal pairs of an opposition. The case of Tümpisa Shoshone (§3.113) is unique in that it was actually possible to search for the existence of such a pair to check whether ʔ is phonemic in the language (i.e., whether there is a word that depends on the existence of ʔ, and nothing else, to be

distinguished from other words). The only minimal pairs found are listed in (2.3).

(2.3) Minimal pairs of ʔ/Ø in Tümpisa Shoshone (Dayley 1989a):

a.	toi	'go to the toilet (from Engl. *toilet*)'	vs.
	toʔi	'pipe'	
b.	moʔo	'hand, fist'	vs.
	-moo-	'decimal (siːmootɨn '10', wahamootɨn '20', etc.)'	

In the complete absence of minimal pairs the potential for homophony is non-existent, but as mentioned above, few grammars provide such details.

2. *Distributional properties*: if a distinction between two phonemes is obliterated, but the two never surface in the same contexts in the language, obliteration of meaning distinction is avoided. Take for example the gliding of word-final p to w in Lama (§3.63), which is classified as neutralizing because there is no way to tell which phoneme is represented by word-final w's in the language (2.4).

(2.4) p - w alternation in Lama (Ourso & Ulrich 1990: 136):

a.	jap-ə	'to buy'
b.	jaw	'buy!'

Potentially, a word-final w in Lama could represent either the p or the w phoneme, but what if phonemic w is never attested word-finally? In such a case obliteration of meaning distinction is avoided because all surface word-final w's must represent the p phoneme. Since Ourso & Ulrich do not provide contextual information it is not possible to determine whether the loss of word-final distinction between the p and w phonemes results in any obliteration of lexical distinctions.

3. *Shift of contrast*: there are several cases in the corpus where the phonetic distinction between two phonemes is obliterated, but contrast is maintained by shifting elsewhere. E.g., in Newari (§3.87) where intervocalic h is deleted but its presence is still manifested through the breathiness of the following vowel. In formal terms, the contrast represented by the h/Ø opposition is not obliterated with the loss of intervocalic h, but simply shifts to the second vowel in the resulting cluster. Although such details are often present in the more comprehensive descriptions of grammars, this is not always the case.

4. *Instrumental Study*: German final devoicing is a process that has been investigated at length and is regularly discussed in literature as an

example of neutralization. In (2.5a) word-final d is devoiced to t, neutralizing the distinction with phonemic t in (2.5b).

(2.5) German Final Devoicing (Hock 1991b: 58):

a.	rāt	'wheel'	rādəs	'of the wheel'
b.	rāt	'advice'	rātəs	'of the advice'

But relatively recent phonetics studies (Mitleb 1981, Port and O'Dell 1985, Port and Crawford 1989) suggest that there are phonetic cues that may keep the devoiced stops distinct from phonemically voiceless sounds in German. This, in turn, suggests that German final devoicing is not an example of complete phonetic neutralization, and therefore does not have the potential to obliterate meaning distinctions. One cannot draw broad conclusions about other languages based on these studies, but they do suggest that even cases that are commonly considered neutralizing may in fact avoid actual obliteration of distinction.

Even when homophony does result from a phonologically neutralizing process, the functional consequences of the obliteration of lexical distinction may be minimal for the following reasons:

1. *Low functional load*: if very few lexical items depend on an opposition to be distinguished from others, the neutralization of this opposition causes little confusion (Bybee 1994). E.g., in Blackfoot (§3.15) w j → Ø / C_, which neutralizes the phonetic distinction w/j/Ø, but is limited to two suffixes (-wa and -ji, both relate to obligatorily possessed nouns) and is additionally restricted by the preceding sounds. Few grammars provide detailed information on the functional load of oppositions, and it is quite possible that some of the processes classified as neutralizing in the corpus may in fact result in little or no obliteration of meaning distinctions, as detailed here. Note also that the existence of minimal pairs and the functional load of an opposition are interrelated: to be truly dependent on an opposition to be distinguished from other words implies membership in such a pair.
2. *Function words*: these depend on fewer acoustic cues for recognition (Lieberman 1963, Gregory 2001, Jurafsky et al. 2001, Bell et al. 2003) and may therefore undergo changes with fewer functional consequences. E.g., in English the suffix s/z marks possessive, plural, and 3rd person, creating homophony, but this is not expected to lead to any miscommunication. Some processes classified as neutralizing in the corpus may also be restricted to function words, or only lead to

homophony of such words, which would entail only negligible consequences for communication.

For these reasons it is quite possible that additional information about a language, information that goes into extensive details beyond the realm of most grammars, may reveal that some of the processes classified as neutralizing do not in fact obliterate any actual meaning distinctions. However, for the sake of the statistical analysis in the present study I rely wholly on available information and equate all cases of *potential* phonological neutralization with *actual* neutralization.

Two more items are added to the summary at the head of the section for each language: trends and differences. Under *trends* I keep track of important tendencies in the corpus in order to observe emerging patterns. The two main trends I keep track of are cases of phonetic, but not phonological, neutralization and cases where flapping patterns with spirantization. Under *differences* I mention if the process in question is changed in any way from the way it appears in Kirchner's inventory (e.g., condensing related processes into one, correcting the formalizations to more closely reflect their descriptions in the original grammar sources, and omitting unattested processes).

2.1.3.2. Consonant Inventories

This is the first section in the body of each language's analysis. The inventory of consonants is presented and identified as either phonemic or not, or left unspecified when this information is not clear. The inventory for each language is transcribed using the IPA (International Phonetic Alphabet) standard and presented in a table that is organized uniformly throughout. The position of every recurring inventoried sound in the 153 languages of the study is illustrated in Tables 3-6. The sounds are split into relevant groups for the sake of clarity.

The columns of each table represent places of articulation from bilabial to glottal. The rows represent manner of articulation, and range from stops to glides, with the voiceless sounds always preceding their voiced counterparts. Table 3 illustrates the position of every inventoried plosive. The order in which similar sounds are presented is often by decreasing simplicity, where p is presented above $p^{ʔ}$ and p^{w}, and $p^{wʔ}$ is presented below them, but sometimes it is also arbitrary (e.g., aspirated stops are followed by glottals, followed by ejectives, followed by labialized). However, all decisions—logical or arbitrary—are consistent throughout.

Table 3: Recurring Plosives in the Inventoried Languages

bilabial	labiodental	dental	alveolar	post-alveolar	retroflex	palatal	velar	uvular	pharyngeal	glottal
p p^j		t̪	t t^j tl ts ts^j	tʃ	ʈ	c cç	k k^j kp	q		ʔ
pː			tː tsː				kː			
p^h		t̪^{h}	t^h tl^h ts^h	tʃ^h	ʈ^h	c^h cç^h	k^h	q^h		
p^ʔ		t̪^{ʔ}	t^ʔ tl^ʔ ts^ʔ	tʃ^ʔ	ʈ^ʔ	c^ʔ	k^ʔ	q^ʔ		
p’			t’ tl’ ts’	tʃ’		c’ cç’	k’	q’		
			t^w ts^w	tʃ^w		c^w	k^w	q^w		ʔ^w
			t^{hw} ts^{hw}				k^{hw}	q^{hw}		
p^{wʔ}		t̪^{wʔ}					k^{wʔ}	q^{wʔ}		
			t’^{w} ts’^{w}				k’^{w}	q’^{w}		
b b^j		d̪	d d^j dl dz	dʒ	ɖ	ɟ ɟʝ	g g^j gb	ɢ		
b^ɦ		d̪^{ɦ}	d^ɦ dz^ɦ	dʒ^ɦ	ɖ^ɦ	ɟ^ɦ	g^ɦ			
b^ʔ										
						ɟʝ’				
			d^w	dʒ^w			g^w	ɢ^w		
^{m}b		^{n}d̪	^{n}d ^{n}dz			^{ɲ}ɟ ^{ɲ}ɟʝ	^{ŋ}g ^{ŋ}gb			
							^{ŋ}g^w			
ɓ			ɗ							

Table 4 illustrates the position of every inventoried spirant. In the inventory chart of each language the rows of spirants follow the rows of stops. As before, sometimes the organization is arbitrary (e.g., ʃ^ʔ before ʃ’) but it is always consistent.

Table 4: Recurring Spirants in the Inventoried Languages

ϕ	f	f^j	s̪	θ	s	s^j	ʃ	ʂ	ç	x	x^j	χ	ħ	h	h^j
					sː										
					s^ʔ		ʃ^ʔ					χ^ʔ			
	f’						ʃ’								
ϕ^w					s^w		ʃ^w		ç^ɣ	x^w		χ^w			
β	v	v^j		ð	z	z^j	ʒ	ʐ	ʝ	ɣ		ʁ	ʕ	ɦ	
					z^w		ʒ^w			ɣ^w		ʁ^w			

Table 5 illustrates the position of every inventoried nasal. In the inventory chart of each language the rows of nasals immediately follow the rows of spirants.

Table 5: Recurring Nasals in the Inventoried Languages

m	mʲ	n̪	n	nʲ	ɳ	ɲ	ŋ	ŋ͡m	ɴ
mː			nː						
mʰ			nʰ			ɲʰ	ŋʰ		
mˀ		n̪ˀ	nˀ						
mʷ			nʷ				ŋʷ		

Table 6 illustrates the position of every inventoried oral sonorant. In the inventory chart of each language the rows of sonorants immediately follow the rows of nasals. The choice of liquids at the top and glides at the bottom is standard, but the order of rows among the flaps, trills, and laterals is arbitrary (and, as before, consistent throughout).

Table 6: Recurring Sonorants in the Inventoried Languages

				ɾ			ɽ			
							ɽʰ			
				r	rʲ					ʀ
				rː						
				rʰ						
			l̪	l	lʲ	ɮ	ɭ	ʎ		
				lː						
				lʰ						
			l̪ˀ	lˀ						
			ɬ̪	ɬ						
				ɬˀ						
				ɺ						
w	ɥ	ʋ		ɹ			ɻ	j	ɰ	
							ɻʰ			
wˀ								jˀ		

Table 7 is a list of sounds that are unique to one language, and do not recur throughout the corpus. They are listed separately because they do not have a standard position in the chart above. In this table they are divided mostly by place of articulation. In the inventory charts of the individual languages they are placed in as logical a manner as possible.

Table 7: Additional (Non-Recurring) Sounds in the Inventoried Languages

pf, bv	ⁿǀ	ʃʷˀ, ʃʲ	ɽ^ɽ	kˣ, kxʰ	qχˀ
pʃʰ, pʃ’, bʒ, fʃ	ɗʷ	ˀd, ˀn	tç, ʈʂ, ʈʂʰ,	ɤ	qᵡ
ˀb, ˀm	ð̞, ðˤ	lʷ	ɕ, ʑ		ħ
ɓʷ	zˤ	l, l̥			

	β̞	ɾ'
		j̞

The sound inventories for each language are collected from the cited grammars, either from similar such tables in the grammars or from descriptions in their text. The modification of this data is mostly limited to the way it is organized into the inventory tables and its translation into IPA, both of which provide consistency to the study and help bring out tendencies.

In a few rare cases it is not absolutely clear which IPA symbol should be used to transcribe a sound. The most common problem surfaced with the symbol c, which in IPA is used to transcribe the palatal stop but is also commonly used to transcribe an alveolar affricate (IPA ts) in other methods. The criteria for the decision in each case remain consistent, and all such judgment calls are recorded in the summary of the languages in which they are made. Any relevance to functional considerations of possible transcription errors are also discussed in the summaries of the individual languages.

2.1.3.3. Descriptions

In this section of the analysis for each language the processes are described as closely as possible to the way they are presented in the grammar sources. Whenever possible examples are included; these too are transcribed in IPA. In order to maintain consistency, language data inside quotations is also translated into IPA.

2.1.3.4. Discussions

In this section of each language's analysis each process is classified as either neutralizing or not, and any issues relevant to this decision are discussed.

2.1.3.5. Additional Comments

In this section additional discussion of interesting trends in the grammar which may not be crucial to determining the functional consequences of the inventoried processes, but are nevertheless somehow relevant, are provided. This section is only included when it is needed and is not present in the analyses of all the languages.

2.1.3.6. Notes on Differences from Kirchner's List

Like the previous section, this is included only when it is necessary. In this section the differences between the way a process is listed in Kirchner's (1998) inventory and the way it is described here, if such differences exist, are detailed. Care is taken to omit processes that are not attested in the sources from which they are cited, to correct the description of processes to more closely resemble the way they are presented in the grammars, and to regroup processes by consistent criteria in order to maintain the legitimacy of the statistics. All these, including the criteria used for each major revision in the way processes are presented, are detailed in this final section of each language.

2.2. Results

2.2.1. Overall Results

If functional considerations did not affect phonetic changes, the number of neutralizing and non-neutralizing processes in the corpus should be fairly even, at or around 50% each.[2] Instead, there is an overwhelming tendency for these phenomena to avoid phonological neutralization, which suggests that phonetic changes such as lenition do not operate independently of functional considerations. The results of the study identifying the functional consequences of 230 phonetically conditioned processes are: 212 processes, which is 92% of the total, are found to be not neutralizing and 18 processes, 8% of the total, are found to be neutralizing (Figure 1).

[2] If sound substitutions were random, since there are more sounds among the world's languages that are not phonemic in a given language than those that are, we would expect substitutions of a phoneme in a language with a non-phonemic sound to be more frequent. This, in turn, would suggest that the avoidance of neutralization may actually be expected to be more common, at about 65% of the time. However, since sound substitutions are not random (i.e., changes such as p→nʲ are unattested, while p→b are rather common) such statistics are not incorporated into the discussion in the present study.

Figure 1: Final Ratio of 230 Processes in 153 Languages

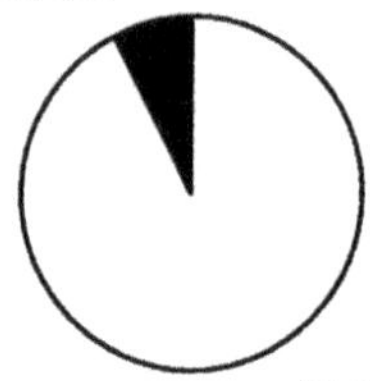

The test for independence in Table 8 establishes that the statistical results are significant. In simple terms, the test for independence calculates the chance that if the sample is increased the results will end up at 50% each. The probability that this would occur, that is, if more processes are added so many of them would be neutralizing that the ratio between neutralizing and not neutralizing processes would even out to 1:1, is 1.8×10^{-37}. This chance is so small that the results of the survey can confidently be considered representative of a larger set of data.

Table 8: Test for Independence

Chi-test Result: 1.8×10^{-37} (=0.0000000000000000000000000000000000018%); p=5%; Sample=230; Actual Range=(212, 18); Expected Range=(115, 115); χ^2=162.9307 with 1 degree of freedom;

Another way to examine how closely the surveyed corpus represents a more general phenomenon is to look at the variation in the statistical results in response to the increase in sampled data. For this test to be significant it is important that the corpus not be sampled in a linguistically-significant order. The order in which the languages, and the processes they exhibit, are presented in Chapter 3 is alphabetical, which from the linguistic standpoint can be considered random. This same order is applied to the statistical test of the ratio distribution of neutralizing to not neutralizing processes as more processes are examined (Figure 2).

Figure 2: Ratio Distribution as Sample Increases

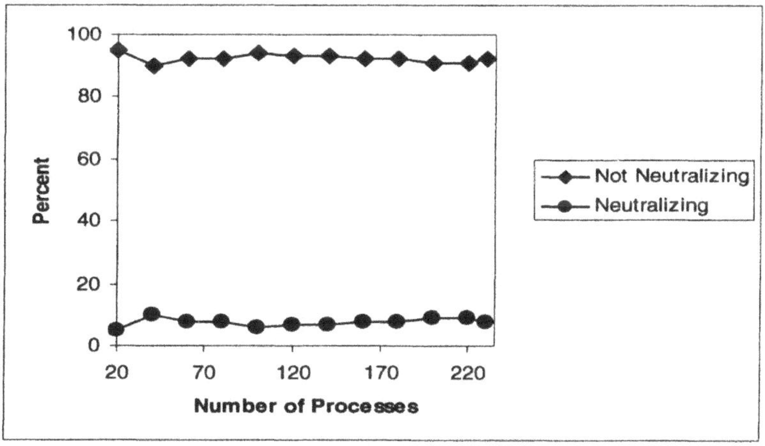

Figure 2 shows that the distribution between neutralizing and not neutralizing processes in the corpus shows little variation, remaining at about 90% not neutralizing to 10% neutralizing, as the number of processes surveyed increases. This confirms the results of the test for independence, that a larger set of data is unlikely to change the final results of the present study. In other words, the overwhelming tendency for phonetically conditioned processes to avoid neutralization cannot be dismissed as an artifact of the sampled inventory, but instead can be viewed as representative of the general tendency of such processes.

2.2.2. Phyla Representation

Although it is unlikely that the lineage of a language would affect the functional consequences of its phonetically conditioned sound changes, in order for results to be considered representative of a general tendency in linguistics they should include a wide variety of language families. In the field of linguistics there is a bias towards the study of Indo European languages, and indeed, much more has been published about these languages than about any other family. However, Kirchner's (1998) inventory comprises a fairly general representation of the world's languages.

Figure 3 illustrates the distribution of the world's languages into phyla, as cataloged on SIL International's online Ethnologue. In all, the Ethnologue lists 105 phyla that include a total of 7029 languages but, for the sake of readability, only the larger families with 1% or more of the world's languages are listed here.

Figure 3: Distribution of the World's Languages into Phyla

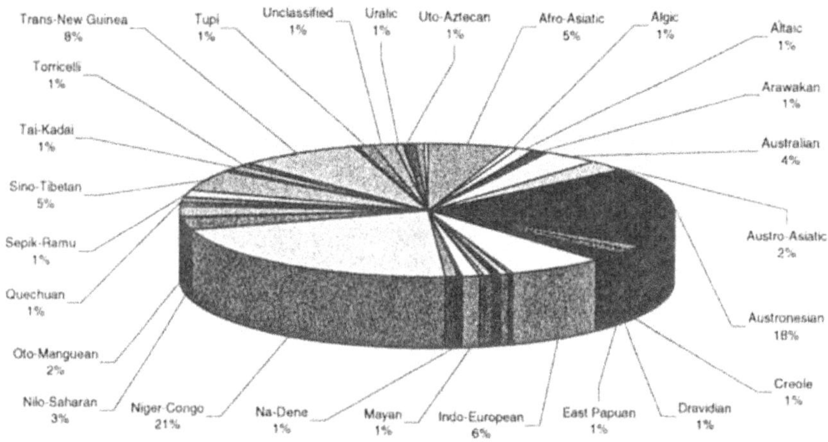

Figure 4 presents a similar look at the 230 processes inventoried in the present study, illustrating the representation of language families in the corpus. This representation within the corpus is not perfect. For example, while Indo European makes up only 6% of the world's languages, it accounts for 28% of the processes inventoried here. Similarly, while Niger-Congo, at 21% of the world's languages includes more languages than any other family, it only accounts for 4% of the inventoried processes. Nevertheless, there is ample representation of a wide variety of language families.

Figure 4: Phyla Distribution of Processes used in Study

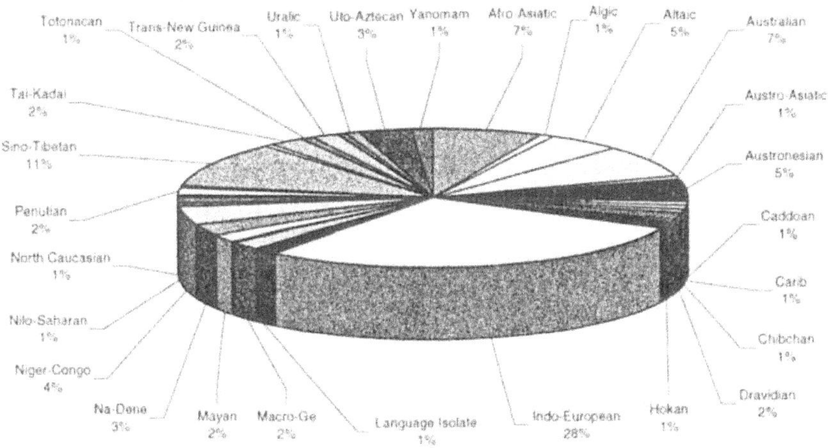

2.2.3. Distribution by Process Type

A closer look at the types of processes of the 230 sampled, and their functional consequences, reveals emerging trends. Figure 5 illustrates the distribution of the 212 non-neutralizing processes into types, and Figure 6 illustrates the same for the 18 neutralizing processes.

Figure 5: Type Distribution of 212 Non Neutralizing Processes

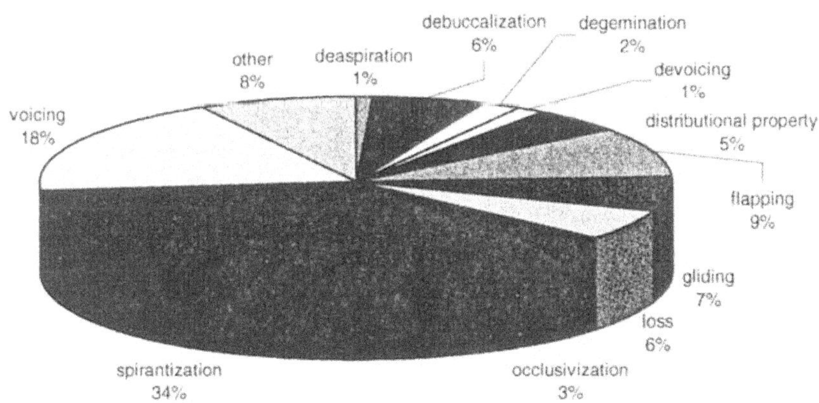

Figure 6: Type Distribution of 18 Neutralizing Processes

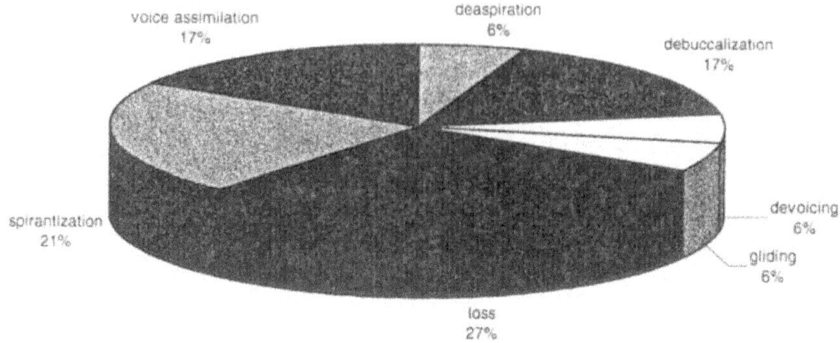

In the case of non-neutralizing processes, 8% are listed as 'other'. This category includes 16 processes that are exclusively non-neutralizing in the corpus; these are listed in Table 9 along with the languages they occur in.

Table 9: 'Other' Non-Neutralizing Processes

Process Type	*Language*	*Process*
Delateralization	Sotho (Southern)	l - d
Dialectal variation	Modern Irish	tense/lax
Glottalization	Lahaul (Paṭṭani)	b d g - bˀ dˀ gˀ
	Maidu	p t k - pˀ tˀ kˀ
Pressure	Arabic (Egyptian)	onset fortis, coda lenis
Lateralization	Limbu (Phedāppe)	t - ʔl (ʔtl)
Laxing	Nepali	b d - 'lax' b d
Preaspiration	Icelandic	pp tt kk - ʰp ʰt ʰk
Shortening	Gojri	duration of geminate
	Thai	m n ŋ w j shorten
Stop release	Laotian (Vientiane)	stops unreleased
Unclear	Arabic (Saʕi:di Egyptian)	fortis - lenis
	Kagate	ɡ - lenis
Unrelease	Lahaul (Paṭṭani)	stops unreleased
	Shapsug (Düzce)	articulatory force varies
Vocalization	Laotian (Vientiane)	w j - u i

At 1 or 2 cases apiece, these types comprise less than 1% each of the non-neutralizing processes and therefore cannot be viewed as indicative of more general trends among the worlds' languages. Note that voicing assimilation in the corpus is also rare, limited to only 3 such cases, but is discussed at some length in the following section. However, the three voicing assimilation cases belong to the 18 neutralizing processes in the corpus, and as such, at 16% of these processes comprise a slightly more significant subsection (see Figure 6).

Figure 7 is a summary of process types. Below the chart the number of processes for each type is recorded, and the figure itself is meant to visually divide each number into neutralizing vs. not-neutralizing processes, so that the ratio of these numbers can be compared.

Figure 7: Summary and Comparison of all Process Types

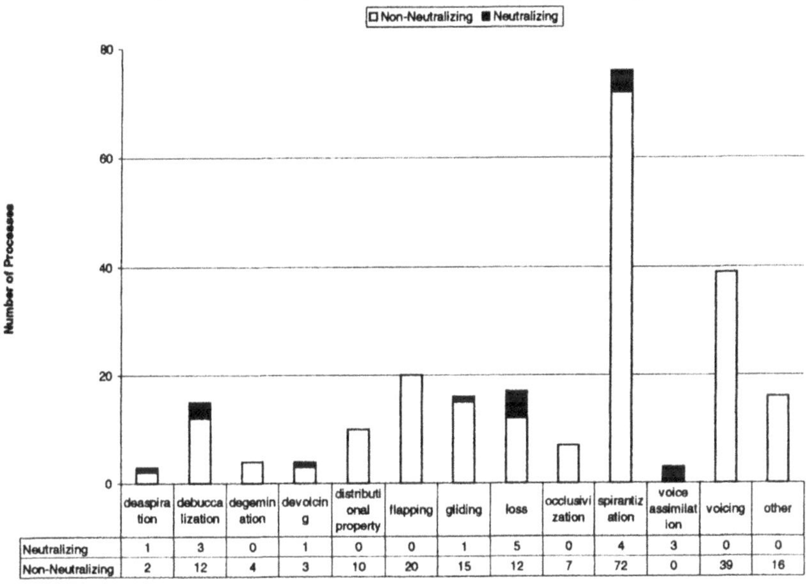

	deaspiration	debuccalization	degemination	devoicing	distributional property	flapping	gliding	loss	occlusivization	spirantization	voice assimilation	voicing	other
Neutralizing	1	3	0	1	0	0	1	5	0	4	3	0	0
Non-Neutralizing	2	12	4	3	10	20	15	12	7	72	0	39	16

The more striking of the trends illustrated in Figure 7 are:

1. *Same functional consequence*: there are several cases where each occurrence of a type of process in the corpus has the same functional consequence. Five types of processes in the inventory are 100% not-neutralizing: degemination (4 cases), DISTRIBUTIONAL PROPERTY, which is a statement about the (usually limited) context in which certain sounds can appear in a language (10 cases), occlusivization (7 cases), flapping (20 cases), and voicing (39 cases). Only one type of process is 100% neutralizing in the corpus: voicing assimilation (3 cases).
2. *The case of spirantization*: spirantization is the most common process in the corpus, at 76 of the 230 processes (or 33% of the total). Of these 5% (4 processes) are neutralizing, and 95% (72 processes) are not.

3. *The case of loss*: there are 17 cases of loss in the corpus, 5 of which are neutralizing and 12 are not. Cases of loss comprise the largest percentage of the neutralizing processes (at 27% of the 18; see Figure 6) but overall more cases of loss in the corpus actually avoid neutralization (71% of the 17 cases of loss in the corpus are not neutralizing).

These trends, and the extent to which they may represent general linguistic tendencies, are discussed in the following section.

2.3. Discussion

2.3.1. The Nature of Neutralizing Processes in the Corpus

In the inventory of 230 processes only 18 are found to be neutralizing. Although this number is too small to qualify for significant statistical analyses of neutralizing processes, some trends can be observed. In §2.3.1.1 the type of process that neutralizes distinction in every case where it applies in the corpus (voicing assimilation) is discussed, and the extent of potential neutralization is addressed in §2.3.1.2.

2.3.1.1. Voicing Assimilation and Pre-Consonantal Position

All the cases of voicing assimilation in the corpus are neutralizing (see Figure 7). Although there are only three such cases, they account for 16% of the neutralizing processes. VOICING ASSIMILATION is a phenomenon where sounds are either voiced or devoiced, depending on the context, to match the voicing value of another sound in the vicinity. For example, in Russian (§3.104) obstruents assimilate in voicing to the following obstruent, which can result in devoicing (2.6a) or voicing (2.6b).

(2.6) Voicing assimilation in Russian (Wade 1992: 12-13):

a.	v xodit	[ˈfxodjɪt]	'goes in'
b.	k domu	[ˈgdomu]	'towards the house'

All three of the cases of voicing assimilation in the corpus are attested in Slavic languages (Bulgarian §3.17, Russian §3.104 and Serbo-Croatian §3.110). It is not surprising that Slavic voicing assimilation is represented 3 times in an inventory of phonetically conditioned processes because it is a phenomenon that has been extensively studied and has received a lot of exposure in literature. But the exclusive lineage of this type of process in the corpus should not be viewed as indicative of more general tendencies among languages.

The three cases of voicing assimilation in the corpus have something else in common: all three affect a distinction in pre-consonantal position. In fact, most of the 18 processes found to be neutralizing affect distinctions in pre-consonantal, or more significantly, non pre-vocalic contexts.

Figure 8: Context of the 18 Neutralizing Processes

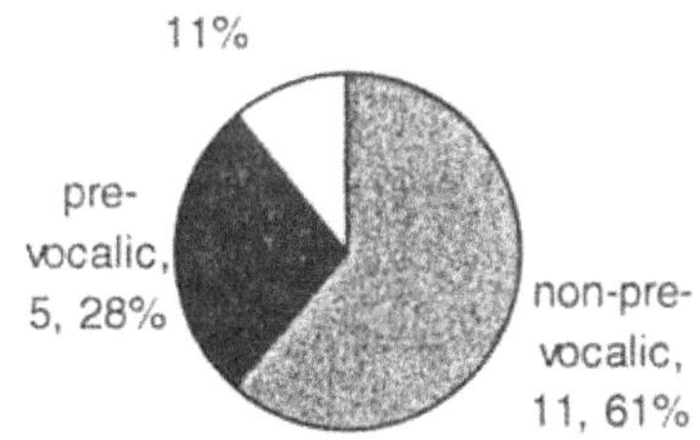

Figure 8 shows that 61% of the neutralizing processes in the corpus operate exclusively in non pre-vocalic positions, 28% exclusively in pre-vocalic positions, and the remaining 11% operate in both contexts (although both these processes are more common in the non pre-vocalic context). While the results here, especially based on such few neutralizing processes (only 18), are certainly not significant enough to warrant broad generalizations on this topic, the trend does confirm the intuition that the non pre-vocalic position is a more likely context for neutralization than pre-vocalic. This intuition is based on work that suggests pre-consonantal distinctions are less perceptible, and harder to maintain, than pre-vocalic distinctions (e.g., Steriade 1999, Ségéral & Scheer 1999).

If contrast considerations indeed constrain the progress of phonetically conditioned processes, then when such processes affect a context that is somewhat less significant for the purposes of contrast maintenance, the resulting changes can be expected to be wider-spread (all other factors being equal, of course). In theory, if pre-consonantal distinctions are less perceptible, fewer words should depend exclusively on them for their distinction, and in turn, neutralization of said distinction should interfere with communication to a lesser extent than the obliteration of a distinction in a pre-vocalic, or onset, position. This would predict that pre-consonantal neutralizations are more numerous than neutralizations in pre-vocalic positions. Although this is true in the present study, where more neutralizations occur exclusively in non pre-vocalic than exclusively in pre-vocalic contexts, the number of neutralizing processes in the corpus is too low to count the pre-consonantal to pre-vocalic ratio as significant. Further research on this issue is warranted.

2.3.1.2. Extent of Neutralization

In Figure 9, the 18 neutralization processes in the corpus are broken down according to factors relevant to the extent of actual neutralization associated with these processes.

Figure 9: Additional Breakdown of the Neutralization Processes

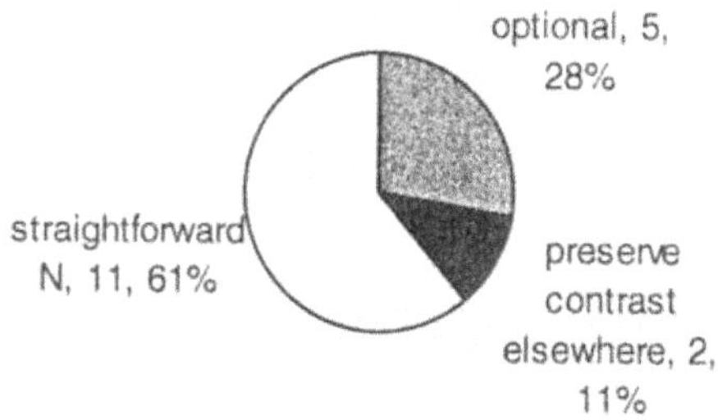

Of the 18 neutralizing processes in the inventory 5 are optional (2 of which are classified as incomplete neutralization), and 2 help preserve contrast elsewhere in the language. Although the 5 optional processes do neutralize contrast in the languages in which they operate, the potential extent of homophony is probably limited by the fact that the offending processes are not compulsory, and in two cases also incomplete. Thus in theory, if the application of a rule may obliterate the distinction between members of an opposition with a high functional load, loss of meaning distinctions can be avoided by merely choosing to ignore the optional rule in question. Such a position, however, is difficult to support without further study into the motivating factors behind the optional aspect of the rules in question as well as more general conditions of optional rule application.

The two processes which serve to preserve contrast elsewhere occur in Toba Batak and Tojolabal. In Toba Batak (§3.133) pre-consonantal p t k debuccalize to ʔ. While this neutralizes the contrast between pp, tp and kp, all of which result in ʔp, by interaction with other processes in the language (namely, n-assimilation, denasalization, and h-assimilation) this glottal formation rule keeps clusters such as pp distinct from np and mh, and clusters such as pt distinct from th (Hayes 1986b). In Tojolabal (§3.134) h j w are deleted word-finally, leading to neutralization among these 3 phonemes. However, the deletion of w actually maintains contrast with post-vocalic word-final u's which, in this very context, surface as w's. Note however, that it is difficult to judge from available data whether the preserved contrasts have a higher functional load than those obliterated, or even whether there is an actual relationship between the distinction-obliterating processes and the resulting maintenance of distinction.

Figure 9 shows that 61% of the neutralizing processes in the corpus are straightforward neutralizations of distinctions between phonemes with the potential to result in widespread obliteration of meaning distinctions, while 39% are somehow limited by either their optional application or the fact that in neutralizing one distinction they also help maintain another. More general implications from statistics based on only 18 processes should be avoided, but these trends deserve a closer look.

2.3.2. The Nature of Non-Neutralizing Processes in the Corpus

2.3.2.1. Types of Processes that are Never Neutralizing

One of the prominent results illustrated in Figure 7 is the five types of processes that, out of the 230 processes investigated, are exclusively non-neutralizing. These are degemination, distributional property, occlusivization, flapping, and voicing.

Degemination is the shortening of a CC cluster, where both consonants are the same, to a singleton C. In languages where geminates are phonemic the CC/C opposition is then in danger of neutralization. However, none of the cases of degemination in the corpus lead to obliteration of meaning distinctions. Consider for example the Uto-Astecan language family Numic (§3.91), where intervocalic *kk shortened to k. Intervocalically, the contrast previously embodied in the Numic opposition *kk/*k is preserved in Tubatulabal as k/h and in Luiseño as k/x.

Table 10: Degemination in the Corpus

Language	*Process*
Bulgarian	CC - C
Numic	*kk > k
Russian	CC - C
Hebrew (Tiberian)	CC - C

There are 4 cases of degemination in all (listed in Table 10), and these comprise 2% of the 212 non-neutralizing processes in the corpus. Since the ratio of non-neutralizing to neutralizing processes in the inventory is 9:1, only 4 non-neutralizing cases of a particular type do not indicate a general tendency of this type of process to always avoid neutralization. Thus, the fact that all 4 degeminations in the corpus avoid neutralization does not imply that this type of phonetically-conditioned change is especially predisposed against obliteration of meaning distinctions. Due to its limited representation in the corpus, therefore, no practical conclusions can be drawn regarding this type of process.

There are 10 cases of distributional properties in the corpus, none of which result in neutralization (Table 11). Again, at a ratio of 9:1 non neutralizing to neutralizing processes, a count of 10 cases on the non neutralizing side is hardly a departure from the expected. However, in this particular case it is significant not because of the number of processes that participate in this category, but because this type of process does not have the potential to obliterate meaning distinction because it does not involve any active sound changes.

Table 11: Distributional Properties in the Corpus

Language	*Process*
Assamese	ɸ β only _#
Badimaya	velars rarely voice
Laotian (Vientiane)	C^h, voiced C only in onset
Panyjima (Mijaranypa)	no r / _#
Punjabi	C: only after central V
Purki	asymmetry in C inventory
Tamazight Berber (Ayt Ayache)	spirantization blocked in C:
Tigrinya	spirantization blocked in C:
Wiyot	no C^h / _#
Yakut	no vd geminate obstruents

A distributional property of a language may well be an indication of some phonetically conditioned change that affected the language in the past, but is not always an indication of what that change may have been. For example, in Wiyot (§3.148) there are no word-final aspirated sounds. This seems to suggest that there may have been de-aspiration of stops in word-final position, but that is not the only logical possibility as the lack of word-final aspiration may have also resulted from aspiration word-initially and internally instead of de-aspiration word-finally. Consider, for example, the case of American English where aspiration is also limited to word-initial and internal contexts (in the onset of stressed syllables; e.g., [p^h]at, a[t^h]omic), a synchronic alternation (atom: a[ɾ]om: a[t^h]omic) that does not involve word-final de-aspiration.

Most important is the fact that the grammar of Wiyot does not describe any active phonological processes that de-aspirate stops in any position, so whatever process may have led to the current ban on word-final aspirates in the language has run its course. As such, no phonemes are undergoing changes and surfacing as variations that may be confused with other phonemes, and therefore, no obliteration of contrast between phonemes is possible.

Another example of a distributional property comes from Punjabi (§3.101), in which geminates occur only after centralized vowels. Again, the restriction of geminates to a certain context does not in itself imply that there is gemination in that context (after centralized vowels) or that there is degemination everywhere else in the language, although it is possible that either gemination or degemination were responsible for bringing about this distribution at some point in the history of Punjabi. In any case, no contrasts are affected by this static property of the language and therefore no obliteration of meaning distinctions is possible.

There are 7 cases of occlusivization in the corpus, none of which are neutralizing (Table 12). Occlusivization is simply a change from a non-stop sound to a stop. For example, in Nyawaygi (§3.92) where the post-consonantal trill r is realized as the alveolar stop d and in Saek (§3.105) where a word-initial spirant ɣ is sometimes realized as the stop g. Both examples involve a shift from a sound that has some degree of oral airflow to a stop, which has none.

Table 12: Occlusivization in the Corpus

Language	*Process*
Car Nicobarese	r - d
English (Liverpool)	θ̪ ð̪ - t d
Nyawaygi	r - d
Saek	ɣ ~ g
Tauya	r - t
Vietnamese (Vinh)	ɣ - g
West Tarangan	j w - g dʒ

Occlusivization is usually classified as fortition, which in terms of *naturalness* and *ease of production* is the opposite of lenition. FORTITION, or STRENGTHENING, is a controversial notion in linguistics usually defined as replacing a sound with one that requires a somewhat greater effort of articulation in a certain environment. In his optimality theoretic account of lenition as effort minimization, Kirchner (1998) considers fortition as "lenition blocking": fortition constraints, along with faithfulness, are the counter force to effort minimization constraints (LAZY). In other words, although fortition constraints are not explicitly perceived as contrast preserving, they are seen as working alongside faithfulness constraints (the contrast maintenance aspect of OT) towards a common goal. As such fortition seems to be viewed by Kirchner as distinction-preserving where lenition is distinction-threatening.

All 7 processes listed in Table 12 are presented in Kirchner's inventory as sound shifts from stops to non-stops, which conforms to his definition of lenition. However, the directionality of these 7 sound shifts is reanalyzed in the present study according to their descriptions in the primary grammar sources for the languages in question. Since occlusivization, an example of fortition rather than lenition, is not the type of process Kirchner set out to catalog, the number of occlusivizations in his inventory is understandably limited, and too small to draw conclusions from based merely on statistics.

There are 20 cases of flapping in the corpus, none of which are neutralizing, listed in Table 13 (there are an additional 15 cases of flapping that pattern with spirantization, none of which are neutralizing either, discussed in §2.3.2.2). Flapping is a process where a sound (in the present corpus this includes 1 instance of a lateral, 5 instances of alveolar stops, 9 retroflex stops and 5 trills) is replaced by either an alveolar or a retroflex flap (ɾ or ɽ).[3] The processes of flapping appear to be especially inclined towards avoiding neutralization: the overall ratio of 9:1 non-neutralizing to neutralizing processes in the corpus suggests that out of 20 cases of a particular type of process we should expect at least 1 or 2 to be neutralizing, but no cases of flapping are. Hence, although the statistics involving flapping in the corpus are not excessively significant, it does appear that flapping is somewhat biased against neutralization.

Table 13: Flapping in the Corpus

Language	*Process*
Canela-Krahô	l - ɾ
Djabugay	r - ɾ
English (American)	t d - ɾ
English (New Zealand)	t - ɾ
Finnish	r - ɾ
Gujarati	ɖ - ɽ
Kagate	r - ɾ
Kashmiri	ɖ - ɽ
Lamani	ɖ - ɽ
Marathi (Halabi)	ɖ$ɖ^{h}$ - ɽ$ɽ^{h}$
Nepali	ɖ$ɖ^{ɦ}$ - ɽ$ɽ^{ɦ}$
Panyjima (Mijaranypa)	ʈ - ɽ
Panyjima (Mijaranypa)	r - ɾ or ɹ
Purki	d - ɽ
Sawai	d - ɾ

[3] Most of the grammar sources used for the present study do not distinguish between *taps* and *flaps*. I follow this practice as well.

Shina	r - ɾ
Shoshone (Tümpisa)	t - ɾ
Tsou (Tfuya)[4]	ˀɗ - ˀl
Urdu	ɖ ɖʰ ~ ɽ ɽʰ
Yindjibarndi	ʈ - ɽ

One interesting aspect of flapping processes in the corpus is that the reason none of them are neutralizing is that the flaps (ɾ and ɽ) are never contrastive in the languages in question, so these processes never affect distinctions between phonemes. That is, in the corpus flapping only occurs where it does not have the potential to lead to neutralization. In §2.3.2.2 I discuss the flapping that patterns with spirantization, and the possibility that the outcome of such processes is related to the systems of contrasts in the given languages.

Finally, there are 39 cases of voicing in the corpus, none of which are neutralizing (Table 14). Of the processes that are exclusively non-neutralizing, this category includes the most items. Since 8% of the processes in the inventory are found to be neutralizing, one would expect at least 3 of the voicing processes in the corpus to be the same. The extent to which this figure is significant is not so far reaching to predict that all cases of voicing among the world's languages avoid obliteration of meaning distinctions, but it does suggest that voicing is usually a non-neutralizing process. That is, the results of the present study show that phonetically conditioned processes are overwhelmingly non-neutralizing, and of those processes, voicing is especially unlikely to obliterate meaning distinctions.

Table 14: Voicing in the Corpus

Language	*Process*
Amele (Haia)	b ~ p
Ao (Chungli)	p t k - b d g
Apalai	k - g
Babine	s - z
Badimaya	Stop voicing
Bulgarian	ts x - dz ɣ
Canela-Krahô	p t k - b d g
Cuna	p t k - b d g
Djabugay	p t c k - b d ɟ g
Estonian (Southern)	obstruent voicing

[4] This process is described in Tsou grammar as both flapping and lateralization (Wright 1996).

Faroese	p t c k - b d ɟ g
French (Périgourdin)	t - d
Gitksan	p t ts k k^{w} q - b d dz g g^{w} ɢ
Gooniyandi	p t̪ t ʈ c k - b d̪ d ɖ ɟ g
Italian (Southern)	p t k - b̥ d̥ g̊
Kannada	k t p - g d b
Korean	stops voiced
Lezgian	p k t ts tʃ q - b g d z ʒ ʁ
Limbu (Phedāppe)	p tʃ k - b dʒ g
Limbu (Phedāppe)	p^{h} t^{h} k^{h} - b^{h} d^{h} g^{h}
Lotha	t - d
Lotha	k - g
Macushi	p t s k - b d z g
Malayalam	p t̪ ʈ k - b d̪ ɖ g
Mbabaram	stops voiced
Mohawk (Akwesasne)	t k - d g
Northern Corsican	p t k - b d g
Northern Corsican	f s - v z
Sanuma	p t ts k - b d dz g
Sekani	s ɬ ç x ʍ - z l ʝ ɣ w
Senoufo	p t k - b d g
Shoshone (Tümpisa)	p t c k k^{w} - b d ɟ g g^{w}
Tahltan (Iskut)	C̥ - C
Tamazight Berber (Ayt Ndhir)	$t^{ʕ}t^{ʕ}$ $s^{ʕ}s^{ʕ}$ - $ð^{ʕ}$ $z^{ʕ}$
Totonac (Misantla)	p t ts tʃ k q - b d dz dʒ g ɢ
Urubu-Kaapor	p t k - b d g (ɣ)
Warndarang	p t ʈ c k - b d ɖ ɟ g
Yakut	p - b
Yana	p t ts k partially voice

Why are voicing processes especially predisposed to avoiding obliteration of meaning distinctions? Figures 10 and 11 show two overwhelming tendencies among the 39 voicing processes in the corpus. Figure 10 shows that 85% of the voicing processes in the present study involve changes in pre-vocalic positions, or syllable onsets, and Figure 11 shows that 79% of the voicing processes avoid obliteration of meaning distinctions due to the fact that voicing is not contrastive in the languages in question. Both trends are discussed below.

Figure 10: Breakdown of Voicing by Syllable Context

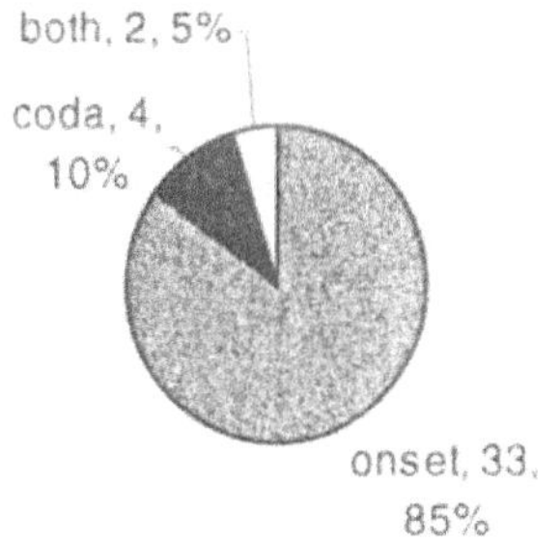

Figure 11: Breakdown of Voicing by Distinction-Maintaining Strategies

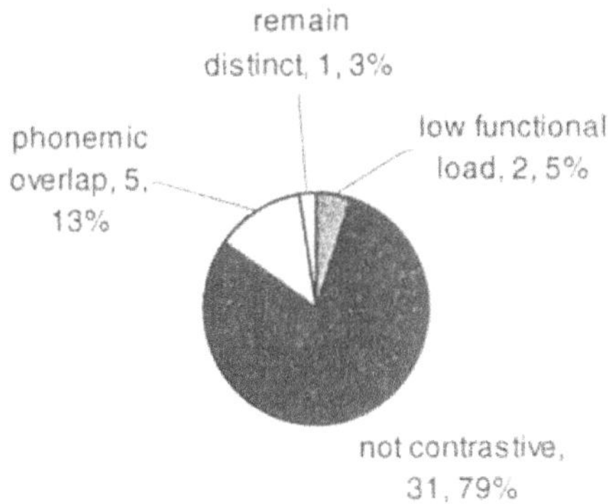

In §2.3.1.1 I discuss the fact that more neutralization is observed exclusively in non pre-vocalic position. The reason given is that pre-consonantal distinctions are less perceptible and therefore more difficult to maintain. Based on this, one would expect that distinctions in pre-vocalic positions would be more perceptible and easier to maintain. One could even go so far as to suggest these distinctions become more crucial to communication, picking up the 'slack' of their pre-consonantal counterparts. With this in mind, processes that affect the onset are predicted to be primarily non-neutralizing. As far as current results go, this prediction is realized.

It is interesting that the most common reason that neutralization is avoided among voicing processes in the corpus is due to the fact that voicing is not contrastive in the languages in which the processes in question operate. In 25 of the 31 cases where voicing is not contrastive this is true for the entire series. E.g., in Djabugay (§3.23) where word-initial p t c k are voiced to b d ɟ g, voicing is not contrastive at all (i.e., the only phonemic stops in the language are p t c k). Thus, no distinctions are affected by the sound shift.

In 6 of the voicing processes in the corpus that avoid neutralization due to lack of contrast, this absence of contrast is limited to the affected sounds or contexts. For example, in Bulgarian (§3.17) ts and x are voiced to dz ɣ, and although voicing in general is contrastive in the language, dz and ɣ are not phonemic. Since voicing is not contrastive for the sounds in question no neutralization between phonemes is possible. By contrast, in Sekani (§3.108) the alternating sounds are all phonemic: the noun/stem-initial fricatives s l̥ ç x ʍ are voiced to z l ʝ ɣ w when prefixed, and all 10 sounds are phonemic in the language. However, the fricatives do not contrast in voicing in the context where the alternation takes place, and therefore, obliteration of distinction between the phonemes is avoided.

The case of voicing provides a unique view into how pervasive a lenition process that does not threaten contrast can become. The bulk of voicing processes in the corpus affect languages where voicing is not contrastive, and most such processes shape entire series (i.e., affect all the stops in a given language, as they do in Djabugay and most other languages in the corpus). As such, the statistics involving voicing processes establish a relationship between the absence of threat to contrast, and the frequency and magnitude of a phonetically conditioned sound change. In §2.3.2.2 I show that spirantization, another common type of lenition, is also far more common where contrast is never threatened. Certainly further research into this matter is necessary before hypotheses about the cause-and-effect aspect of this relationship can be formed, but the correlation is clear.

2.3.2.2. The Case of Spirantization

The most common type of process in the corpus is spirantization, which accounts for 76 of the 230 processes investigated. Of these 72, or 95%, are non-neutralizing. This is the type of process that most closely represents the ratio of 9:1 non-neutralizing to neutralizing processes in the corpus. That is, the over 90% non-neutralizing ratio of spirantization processes is not in itself interesting because, as the overall results of the present study show, this is exactly what we would expect. What is interesting about spirantization is that it is so common, and its incorporation of flapping processes, described below.

One of the primary goals of Kirchner's compilation of lenitions is in fact to show how common spirantization is, thus distinguishing his approach from previous methods as the only one that can account for the naturalness, and hence frequency, of such processes. The overwhelming majority of spirantization cases in the compilation of lenition processes may very well be a consequence of this goal. Nevertheless, there is no denying the fact that, as far as phonetically conditioned processes go, spirantization is indeed common.

In Figure 12 the breakdown of the distinction maintaining strategy of the 72 cases of non-neutralizing spirantization in the corpus is presented. As in the case of voicing in the corpus, which is also a common type of lenition, the reason no meaning distinction is neutralized in most cases is because contrast is never threatened by these processes. That is, in 92% of such processes the outcome of the spirantization results in a sound that is not contrastive in the language, and therefore distinction between phonemes is never affected. This data confirms the generalization made regarding voicing processes in the corpus, that lenitions which do not threaten contrast are more likely to proceed unhindered, be common, and affect entire series of sounds (see §2.3.2.1).

Figure 12: Breakdown of Spirantization by Distinction-Maintaining Strategies

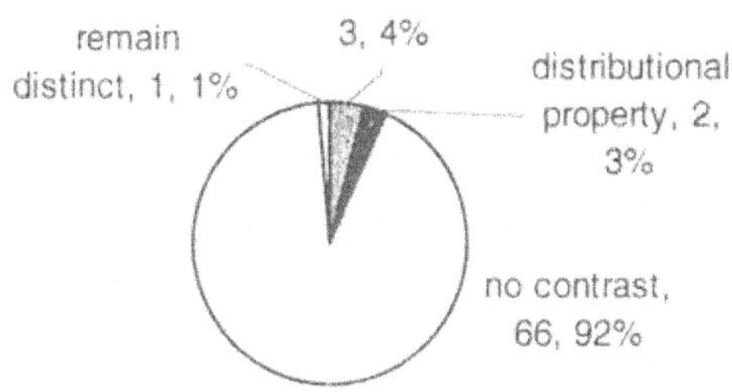

Another interesting tendency among the spirantization phenomena in the corpus is their incorporation of 15 cases of flapping. The processes of flapping that pattern with spirantization in the corpus account for 21% of the total of spirantizing processes (Figure 13), and are never neutralizing.

Figure 13: Non-Neutralizing Spirantization in the Corpus

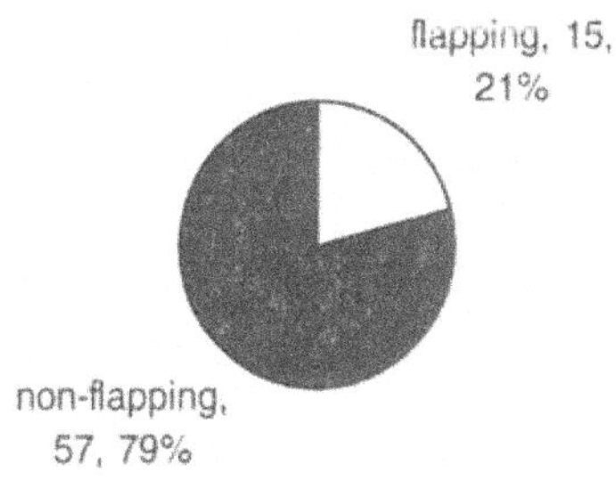

These processes are itemized in Table 15, along with the languages they appear in and whether the flapping affects the alveolar or the retroflex stops. Note that flapping only patterns with spirantization that either involves some

voicing or affects voiced sounds; i.e., ɾ patterns with β and ɣ, but not with ɸ and x.

Table 15: Flapping Processes that Pattern with Spirantization

Language	*Process*	*Alveolar*	*Retroflex*
Efik (Calabar-Creek)	b d k - β ɾ ɣ	X	
Gujarati	bʰ dʰ ɖʰ gʰ - Hβ Hð Hɽ Hɣ		X
Karao	p t d tʃ dʒ gʷ q - ɸ θ l ɾ j w χ	X	
Kupia (Sujanakota)	p ʈ ɖ - ɸ ŧ ɽ		X
Lowland Murut (Timugon: Poros)	b d g - β ɾ ɣ	X	
Malayalam	b d̪ ɖ g - β ð̪ ɽ ɣ		X
Moghamo (Batibo)	d g - ɽ ɣ	X	
Proto-Bantu	b d g - β l ɣ	X	
Sanuma	d dz - ɾ z	X	
Senoufo	b d g - β ɾ ɣ	X	
Shina	b d ɖ g - ƀ đ ɽ ǥ		X
Somali	b d ɖ g - β ð ɽ ɣ		X
Taiwanese	p t k - β ɾ ɣ	X	
Tamil	p t ts ʈ k - β ð dz ɽ ɣ		X
Tharu (Chitwan)	pʰ b ɖ ɖʰ dʒ - ɸ β ɾ ɾʰ ʒ		X

There are 7 languages in which a retroflex stop undergoes flapping (usually to a homorganic flap) as part of a more general spirantization process that affects other stops. For example, in Somali (§3.114) where the intervocalic stops b d g spirantize to β ð ɣ, the intervocalic retroflex stop ɖ is flapped to ɽ (2.7).

(2.7) Spirantization of the Somali stops (Armstrong 1964):

a.	'laba	['laβa]	'two'
b.	'badag	['baðag]	'goose'
c.	'tiɖi	['tiɽi]	'she said'
d.	'sagaal	['saɣaal]	'nine'

In the present study all retroflex stops are flapped where other stops are spirantized. The case of the alveolar is more interesting, as one would expect d to pattern with the rest of the stops and surface as ð where b g → β ɣ. This is indeed the case in the majority of the spirantization processes in the corpus that involve voicing as either part of the process or a feature of the affected sounds (25 cases). However, in 8 languages the alveolar stop flaps as part of this same type of process. In Proto-Bantu (§3.98) where b g spirantize to β ɣ the d is realized as l, but in the other 7 languages the alveolar stops (or in the case of

Karao, affricate) flap to ɾ. E.g., in Taiwanese (§3.123) where intervocalic p k spirantize and voice to β ɣ, intervocalic t is realized as ɾ (2.8).

(2.8) Spirantization in Taiwanese (Hsu 1995: 5):

a.	apa	[aβa]	'little box'
b.	k^{h}uta	[k^{h}uɾa]	'little hole'
c.	tika	[tiɣa]	'bamboo'

As mentioned above, there are 8 languages where the flapping of the alveolar stop patterns with the spirantization of other stops (henceforth dubbed ɾ-languages), where voicing is involved as either part of the process, or a feature of the affected sounds. There are 25 languages where the same type of spirantization process changes the alveolar to ð (ð-languages). Other spirantization processes in the corpus either do not involve voicing or do not affect the alveolar stops. These 33 languages pattern with some regularity: first, none of the spirantization processes in either the 8 ɾ-languages or the 25 ð-languages are neutralizing. Second, the phonemic inventories of most ɾ-languages do not include a trill, while the phonemic inventories of most ð-languages do.

Table 16 provides a list of phonemic inventories of 3 of the 8 ɾ-languages: (a) Calabar-Creek Efik (§3.24), where b d k → β ɾ ɣ non-initially in a word stem and before a vowel; (b) Poros sub dialect of Timugon Lowland Murut (§3.69), where b d g → β ɾ ɣ after vowels, semi-consonants and d g ʔ, and clause-finally; and (c) Sanuma (§3.106), where d dz → ɾ z / V_V in fast speech. Note that not one of these inventories includes a phonemic trill.

Table 16: Inventories of 3 ɾ-Languages

(a) *Efik*

		t			k	kp
b		d				
	f		s			
m		n		ɲ	ŋ	
w				j		

(b) *Lowland Murut*

p	t				k	ʔ
b	d		dʒ		g	
		s				
m	n				ŋ	
	l					
w				j		

(c) *Sanuma*

p	t	ts	k	
	t^{h}			
		s		h
m	n			
	l			
w				

Only one ɾ-language in the corpus appears to pose a possible exception to the absence of the phonemic trill rule: in Karao (§3.55) a trill is listed as part of the consonant inventory. However, the consonant inventory for this language is not limited to phonemes.

In contrast, the 25 ð-languages in the corpus regularly include a phonemic r. The inventories of two languages that follow the phonemic r trend, and the only exception in the study, are provided in Table 17: (a) Eastern Catalan (§3.20), where b d g → β ð ɣ medially; (b) Badimaya (§3.11), where t̪ c → ð ʒ / V_V; and (c) Ata Manobo (§3.73) is the only exception in the corpus, where b d g - β ð ɣ but there is no phonemic trill in the inventory of consonants.[5]

Table 17: Inventories of 3 ð-Languages

(a) *Catalan*

p		t	ts	tʃ		k
b		d	dz	dʒ		g
	f		s	ʃ		
			z	ʒ		
m		n			ɲ	
		ɾ/r				
		l			ʎ	

(b) *Badimaya*

p	t̪	t	ʈ	c	k
m	n̪	n	ɳ	ɲ	ŋ
			ɽ		
		r			
	l̪	l	ɭ	ʎ	
w				j	

(c) *Manobo*

p	t			k	q	
b	d			g		
		s				h
m	n			ŋ		
	l					
w			j			

The trend of the alveolar flapping where other stops spirantize in languages which do not include a phonemic trill is highly regular, as is the trend of d → ð in languages where r is phonemic. I propose that the connection is one of contrast maintenance: for alveolars, flapping may be the preferred outcome of the process that spirantizes the rest of the stop series, as is the case for the retroflex stops which always flap where other stops spirantize. But flapping may be avoided when the threat of contrast neutralization is present, which appears in the form of a phonemic trill, a sound that is phonetically similar to a flap[6], and in such cases an alveolar stop is spirantized to ð. This is supported by the fact that absolutely no flapping processes in the corpus ever lead to neutralization (see §2.3.2.1 for a discussion of the 20 cases of flapping that do not pattern with spirantization). Thus, where voicing and spirantization provide evidence of the relationship between the system of contrasts in a language and the extent of

[5] The only grammar source available on the Ata dialect of Manobo is scarce on details; other dialects of Manobo, e.g., Western and Southern Bukidnon Manobo (Elkins 1970), do have phonemic trills and there is reason to believe that the exclusion of r from the phonemic inventory of Ata may be a consequence of the author's analysis. In other words, this language may not be an exception to the trend of ð-languages always including a phonemic trill after all.

[6] Obviously ɾ and r are not identical, and they do contrast in some languages. Of the 153 languages in my corpus, 3 appear to have this contrast: Kandoka-Lusi Kaliai-Kove (§3.51), Mexico City Spanish (§3.121), and Tzeltal (§3.139), which is one of the ð-languages. However, this contrast does appear to be rare.

phonetically conditioned sound changes, the case of flapping suggests a relationship between the system of contrasts in a language and the outcome of such processes.

2.3.2.3. Meaning Distinction Maintaining Strategies

There are 17 cases of deletion, or loss in the corpus, 12 of which are found to be not-neutralizing. This figure is somewhat surprising because, as the ultimate step in lenition processes, loss results in the elimination of a phoneme in some context, obliterating the phonetic distinction between the phoneme in question and Ø. But in 71% of the cases of loss in the corpus this does not result in phonological neutralization: no meaning distinction is actually obliterated.

In fact, there are 52 cases of phonetic, but not phonological, neutralization in the corpus (see §1.0 or the Glossary in Appendix B for the distinctions that are made between these types of neutralization). These are listed in Table 18, along with the phonetic contrasts neutralized as well as the way in which meaning distinctions are maintained in each case. Note that more than one process could lead to the phonetic neutralization of the same contrast in a language, which is why some entries in Table 18 are duplicates. E.g., the b/f contrast is affected in Haia Amele (§3.1) by two separate processes in the corpus: (1) f → p / #_, V_V and (2) b → p / _#.

Table 18: Meaning Distinction Maintenance in the Corpus

Language	*Phonetic Contrasts Neutralized*	*Strategy for Distinction Maintenance*
Amele (Haia)	b/f	varying context
Amele (Haia)	b/f	varying context
Babine	s/z	low functional load
Babine	x/h	phonemic overlap
Blackfoot	w/j/Ø	low functional load
Bulgarian	N/Ø	contrast shifts
Canela-Krahô	j/x	low functional load
English (American)	t/d	contrast shifts
English (Cardiff)	b/p d/t g/k dʒ/tʃ	contrast shifts
English (New Zealand)	t/d	contrast shifts
French (Périgourdin)	t/d	phonemic overlap
French (Périgourdin)	t/d	phonemic overlap
Gbɛya	ʔ/Ø	low functional load
Gujarati	ɖʰ/ɖ	contrast shifts
Haitian Creole	j/Ø	low functional load

Kuna (Paya)	p/b t/d tʃ/dʒ k/g	contrast shifts
Lezgian	p/b t/d ts/z tʃ/ʒ k/g q/ʁ	low functional load
Limbu (Phedāppe)	p/b	phonemic overlap
Limbu (Phedāppe)	t/l	phonemic overlap
Limbu (Phedāppe)	p/b	phonemic overlap
Malayalam	p/b t̪/d̪ ʈ/ɖ k/g	phonemic overlap
Malayalam	p/b t̪/d̪ ʈ/ɖ k/g	phonemic overlap
Mataco (Noctenes)	ʔ/h/Ø	varying context
Moghamo (Batibo)	t/d k/g	phonemic overlap
Navajo	x/h	varying context
Navajo	ɣ/j/Ø	low functional load
Nepali	ts^h/h	phonemic overlap
Nepali	h/Ø	contrast shifts
Newari	h/Ø	contrast shifts
Nez Perce	h/Ø	low functional load
Northern Corsican	p/b t/d k/g	phonemic overlap
Northern Corsican	p/b t/d k/g	phonemic overlap
Northern Corsican	f/v	phonemic overlap
Northern Corsican	f/v	phonemic overlap
Numic	kk/k	phonemic overlap
Sekani	s/z ɬ/l ç/ʝ x/ɣ ʍ/w	varying context
Senoufo	p/b t/d k/g	phonemic overlap
Senoufo	p/b t/d k/g	phonemic overlap
Shapsug (Düzce)	p/b t̪/d̪ t/d c/ɟ k/g	low functional load
Shoshone (Tümpisa)	ʔ/h/Ø	contrast shifts
Southern Tati (Chali)	h/Ø	contrast shifts
Spanish (Mexico City)	p/b t/d k/g	phonemic overlap
Tauya	k/ʔ k^w/$ʔ^w$	varying context
Tauya	r/t	low functional load
Tojolabal	r/g	contrast shifts
Tojolabal	r/g	contrast shifts
Turkish	g/Ø	contrast shifts
Warndarang	p/w c/j	contrast shifts
Yakut	p/b	varying context
Yana	b/w	low functional load
Yindjibarndi	p/w t̪/j ɻ/j k/w	contrast shifts
Yuman	c/s	phonemic overlap

There are four ways in which meaning distinctions are maintained in cases of phonetic neutralization: low functional load, due to varying distributional properties, contrast shifts, and phonemic overlap. The distribution of the 52 processes among these strategies is summarized in Figure 14, and the strategies themselves are detailed below.

Figure 14: Summary of 52 Meaning Distinction Maintenance Strategies

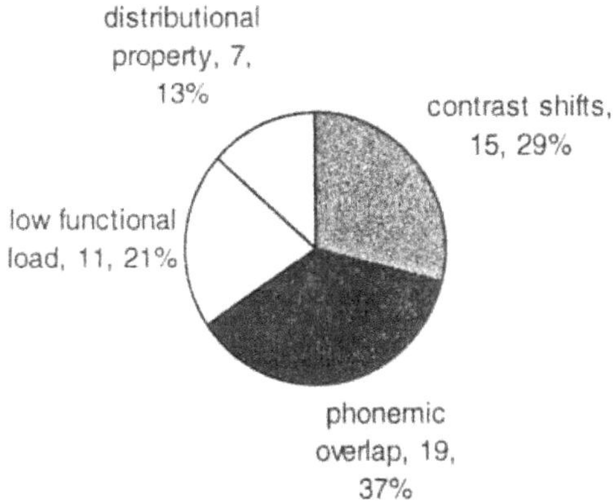

1. *Low Functional Load*: as discussed in §2.1.3.1, when very few words depend on an opposition to be distinguished from each other the neutralization of said opposition causes little confusion. E.g., in Blackfoot (§3.15) where w and j are lost in post-consonantal position, which neutralizes the phonetic distinction between w, j, and Ø, but does not lead to the obliteration of meaning distinctions because the process only affects two suffixes (-wa and -ji, both relate to obligatorily possessed nouns) and is additionally restricted by the preceding sounds. 21% of the maintenance of meaning distinctions in cases of phonetic neutralization is due to this.
2. *Distributional properties*: as discussed in §2.1.3.1, if a phonetic distinction between two phonemes is obliterated but they never surface in the same context, obliteration of meaning distinctions is avoided. For example in Haia Amele (§3.1) the phonetic distinction between b and f is neutralized because both phonemes can be realized as p. However, the realization of phonemic b and f as the non-phonemic p occurs in different environments: only word-finally for b, and only non-finally for f. Thus, it is always clear which phoneme is represented by p, and therefore obliteration of meaning distinctions is avoided. At 13% this is the least common method of contrast maintenance among phonetically neutralizing processes in the corpus.

3. *Contrast shifts*: when the phonetic distinction between two contrastive sounds is neutralized, the contrast previously embodied by the neutralized opposition may shift elsewhere. A common example of this phenomenon is in American English (§3.25) where intervocalic t and d are both realized as ɾ, obliterating the phonetic distinction between the two phonemes. However, the contrast previously embodied by the t/d opposition shifts to the length and height of the preceding vowel. At 29% this is the second most common method of contrast maintenance among phonetically neutralizing processes in the corpus. It is especially prevalent in cases of loss.
4. *Phonemic Overlap*: as discussed in §1.2.2, phonemic overlap is when in some context a phoneme A shifts to B, which is also phonemic in the language, but the A/B opposition is maintained because in the same context B is realized as another sound, C. In §1.2.2 the example from Northern Corsican (§3.90) is presented, where intervocalic p t k are voiced to b d g, and in the same context b d g are spirantized to β ð ɣ. Another example of phonemic overlap is the diachronic case of Cocopa Yuman (§3.153) where *c > s, a phoneme that existed in the language, leading to the phonetic neutralization of the opposition *c/*s. However, *s > ʂ, so the contrast between *c and *s was not obliterated but instead shifted to s/ʂ. At 37% this is the most common method of avoiding obliteration of meaning distinctions among phonetically neutralizing processes in the corpus.

The results of the present study offer a surprising insight: the outcomes of both loss and sound mergers, two phenomena that are phonetically neutralizing and therefore in general expected to obliterate meaning distinctions, in fact rarely do. Furthermore, a closer investigation of such phenomena reveals that meaning distinctions are maintained in one of only 4 ways, 2 of which (phonemic overlap and contrast shift, between which there is not a crucial distinction) can be conflated under a more general heading of 'active sound shifts' totaling 66% of the meaning maintenance strategies. This confirms Silverman's (2000) prediction that phonetically conditioned changes may be accompanied by additional changes motivated by meaning-maintenance consideration: these sound shifts are the most common meaning maintenance strategies, and may very well be such contrast-motivated changes.

2.3.3. Result Summary

The results of the present study show an overwhelming tendency for phonetically conditioned processes to avoid neutralization, an overwhelming tendency for voicing processes in particular to do the same, and an unexpected tendency for loss and other phonetic neutralization to follow suit. Furthermore, a trend that applies to syllable position is identified: processes that affect the non pre-vocalic position are more likely to be neutralizing while processes that affect the pre-vocalic position are less likely to obliterate meaning distinctions.

The strategies for maintaining meaning distinctions in the face of the obliteration of phonetic contrast are detailed as well. Of these, active changes (phonemic overlap and shift of contrast) are the more prevalent, confirming Silverman's (2000) prediction that meaning maintenance may motivate additional changes in response to phonetically conditioned ones.

CHAPTER 3

Languages

3.1. Amele (Haia)

Language: Amele *Dialect*: Haia
Phylum: Trans-New Guinea, Madang-Adelbert Range, Madang, Mabuso, Gum
Inventory: Lavoie *Source*: Roberts 1987
Lenition(s): 1. p ~ f / #_ and V_V (limited)
2. b ~ p / _#
Type: 1. Spirantization
2. Voicing
Classification: 1. Not Neutralization
2. Not Neutralization
Trends: Phonetic, but not phonological, neutralization (b f - p)
Difference: Description differs from Kirchner's

Consonant Inventory

Table 19: Phonemic Consonant Inventory of Haia Amele

		t					ʔ
b		d			g	gb	
	f		s				h
m		n					
		l					
w				j			

Description

Both b and f alternate with p: p ~ f word-initially and intervocalically, but in limited cases; b ~ p word-finally.

Discussion

The author analyzes b and p as phonetic realizations of the phoneme b. p surfaces word finally in polysyllabic words (3.1a) and b surfaces elsewhere (3.1b).

(3.1) Realizations of the phoneme b (p.333):

a.	[hɪˈnup]	'5 days hence'
b.	[gɔb]	'knee'

Roberts also notes that for some lexical items p fluctuates with f word-initially and intervocalically (3.2) and that f and b are contrastive (3.3):

(3.2) Examples of p ~ f alternations word-initially and intervocalically (p.333):

a.	[gæˈpæt]	~	[gæˈfæt]	'canoe'
b.	[ˈpupu]	~	[ˈfufu]	'wind'

(3.3) Contrast between f and b (p.333):

a.	[fæ]	'doubt'	vs.	[bæ]	'today'
b.	[fɛʔ]	'see'	vs.	[bɛʔ]	'come up'

b and f are separate phonemes but both can be realized as p. The contrast between them is maintained because the p ~ f and p ~ b alternations occur in different environments: p ~ b is limited to word-final position while p ~ f occurs only word-initially and intervocalically. Therefore, all word-final p's in the language are manifestations of b while the non-final p's are all manifestations of f, and the contrast between the two phonemes is never obliterated.

Additional Comments

Note that the author considers f to be phonemic and p its phonetic variation. In this case the decision appears to be arbitrary, and does not affect the functional classification of the process as not neutralizing.

Notes on Difference from Kirchner's list

Citing Lavoie (1996), Kirchner (1998) lists the lenition in Amele as p ~ f non-initially. I adhere to the description in Lavoie's grammar source (Roberts 1987) which specifies that the alternations occur word-initially and intervocalically, and only to a limited extent in both contexts.

3.2. Ancient Greek (Classical Attic)

Language:	Ancient Greek	*Dialect*: Classical Attic (& all dialects eventually)
Phylum:	Indo European, Hellenic	
Inventory:	Lavoie	*Source*: Bubeník 1983
Lenition(s):	Voiceless aspirated stops - voiceless fricatives (context-free)	
Type:	Spirantization	
Classification:	Not Neutralization	

Consonant Inventory

Table 20: Phonemic Consonant Inventory of Ancient Greek

p	pː	t		tː	k	kː	
p^h		t^h			k^h		
b		d			g		
			s	sː			h
m	mː	n		nː			
		r		rː			
		l		lː			

Description

Bubeník writes that:

> As shown by Modern Greek, the ancient aspirates /p^h/, /t^h/, /k^h/ changed into fricatives /f/, /θ/, /x/. The beginnings of this process of 'frication' are usually placed in the Greco-Roman period (starting circa 150 B.C.). (p.105)

Discussion

Bubeník describes the frication shift as starting with clusters, where fθ and xθ appeared as allophones of the aspirate p^ht^h and k^ht^h, and eventually the shift was felt in all contexts. The author specifies that contrast between the aspirates and other sounds was maintained in the shift. The fact that no contrast was obliterated is further evident from the inventory that shows no existing spirants with which the results of the frication shift could have neutralized.

3.3. Ao (Chungli)

Name:	Ao	*Dialect*: Chungli
Phylum:	Sino-Tibetan, Tibeto-Burman, Kuki-Chin-Naga, Naga	
Inventory:	Kirchner	*Source*: Gurubaswe-Gowda 1975
Lenition(s):	Stops voice V_V or after voiced C	
Type:	Voicing	
Classification:	Not Neutralization	

Consonant Inventory

Table 21: Phonemic Consonant Inventory of Chungli Ao

p	t			c	k	?
		s				
		z				
m	n				ŋ	
	l		ɮ			
w				j		

Description

The simple stops are voiced intervocalically and when following a voiced consonant.

Discussion

In Ao p t k freely vary with the aspirates p^h t^h k^h in initial position, and are voiced to b d g intervocalically and when following a voiced consonant. Elsewhere they are voiceless unaspirated stops p t k, and are unreleased when followed by a word juncture.

(3.4) Examples of word-initial aspiration (p.10):

a.	pa˩˥	[pa˩˥~p^ha˩˥]	'he'
b.	ti˩˥	[ti˩˥~t^hi˩˥]	'eight'
c.	ka˩˥	[ka˩˥~k^ha˩˥]	'one'

(3.5) Examples of medial voicing (p.11):

a.	apu/|	[abu/|]	'blow'
b.	a\\|tep	[ʌ\\|dep]	'patch'
c.	loŋki	[lɔŋgi]	'hole (made by insect)'

Throughout his transcriptions, except in the section where he describes the aspiration and voicing of the simple stops, Gurubaswe-Gowda uses only p, t and k, and does not transcribe aspiration, voicing, or unreleased counterparts of these sounds.

Based on the fact that b d g appear in predictable environments (intervocalically and after a voiced consonant) and are not phonemic in Ao, I conclude that they are representations of the p t k phonemes. Therefore no contrast is obliterated by the voicing of p t k.

3.4. Apalai

Language: Apalai *Dialect*: Not specified
Phylum: Carib, Northern, East-West Guiana, Wayana-Trio
Inventory: Lavoie *Source*: Koehn & Koehn 1986
Lenition(s): k - g
Type: Voicing
Classification: Not Neutralization
Difference: Corrected context

Consonant Inventory

Table 22: Phonemic Consonant Inventory of Apalai

p	t				k	ʔ
m	n					
		s	ʃ			
		z				
	r					
w				j		

Description

k is voiced to g at morpheme boundaries after elision of a final unstressed vowel and preceding a morpheme initial stressed a.

(3.6) Example (p.121):

jtoko ase [ɨ. ˈtõ. ˈga.se] ‘I am going’

Discussion

There are very few details provided in the source about the voicing process in question so the classification of this lenition process rests on the phonemic inventory. Since there are no voiced velars in the phonemic inventory of this language, the outcome of the voicing process cannot be confused with existing sounds and therefore does not obliterate any contrasts in Apalai.

Notes on Difference from Kirchner’s list

Citing Lavoie (1996), Kirchner (1998) lists the voicing process in Apalai as k → g / _#. However, Lavoie’s source (Koehn & Koehn 1986) describes the context where k voices as occurring at morpheme boundaries after elision of a final unstressed vowel and preceding a morpheme initial stressed a. Datum (3.6) confirms that this process does not occur word-finally.

3.5. Apatani

Language: Apatani *Dialect*: Not specified
Phylum: Sino-Tibetan, Tibeto-Burman, North Assam, Tani
Inventory: Kirchner *Source*: Abraham 1985
Lenition(s): b → β / V_V[+back/central]
Type: Spirantization
Classification: Not Neutralization
Difference: The context of this process varies from Kirchner’s description

Consonant Inventory

Table 23: Phonemic Consonant Inventory of Apatani

p	t	ts		k	
b	d	dz		g	
		s		x	h
m	n			ŋ	
	r				
	l				
			j		

Description

In Apatani b is realized as a semi-voiced bilateral continuant β when it occurs intervocalically, provided the second vowel is a back/central vowel.

(3.7) Example (p.13):

a. aβa 'father'
b. aβuje 'various'

Discussion

Very little information is provided in the source about the spirantization of b to β, so the functional classification of the process is based purely on the phonemic inventory of the language. Abraham does not include β as a phoneme of Apatani which suggests that it is a sound that surfaces in the language only as a result of the spirantization of b, and therefore does not neutralize any contrasts. This is further confirmed by Abraham's use of generativist symbols where virgules ("/b/") identify b as the basic phoneme and square brackets ("[β]") indicate that β is an allophone.

3.6. Arabic (Egyptian)

Language: Arabic *Dialect*: Egyptian
Phylum: Afro-Asiatic, Semitic, Central, South, Arabic
Inventory: Kirchner *Source*: Harrell 1957
Lenition(s): All onset consonants are fortis, coda lenis
Type: Intercostal Pressure
Classification: Not Neutralization

Consonant Inventory

Table 24: Consonant Inventory of Egyptian Arabic

		t				k	(q)		ʔ
b		d				g			
	f		s	ʃ			χ	ħ	h
			z	(ʒ)			ʁ	ʕ	
m		n							
		r							
		l							
w					j				

The emphatic consonants in the language are treated separately from this inventory of sounds.

Description

Harrell notes that all consonants are fortis when releasing a syllable and lenis when closing a syllable.

Discussion

There is very little information about the phonology of this language in the source; the author does not provide examples of sounds undergoing this process or any illustration of the actual difference between a consonant's lenis and fortis articulation. However, Harrell does list this process under "allophonic features" of the language, does not suggest in any way that this fortis/lenis difference is distinctive, or that the fortis/lenis articulation of the same sound results in a different phone. Therefore, regardless of the scarcity of details, this process does appear to be a straightforward case of allophonic variation where no contrast is neutralized.

3.7. Arabic (Saʕi:di Egyptian)

Language: Arabic *Dialect*: Saʕi:di Egyptian
Phylum: Afro-Asiatic, Semitic, Central, South, Arabic
Inventory: Kirchner *Source*: Khalafallah 1969
Lenition(s): All consonants have lenis allophones
Type: Not clear
Classification: Not Neutralization

Consonant Inventory

Table 25: Phonemic Consonant Inventory of Saʕi:di Egyptian Arabic

		t		tʃ		k			ʔ
b		d		dʒ		g			
	f		s	ʃ			χ	ħ	h
			z				ʁ	ʕ	
m		n							
		r							
		l							
w					j				

Description

Very few details are provided about the alternations in question. The author only specifies that all consonants have fortis allophones in post-junctural positions before stressed vowels, and lenis allophones in the environments VC#, where V is unstressed, and VC_#, where V is stressed. No explanation is provided about how the fortis/lenis distinction may be manifested.

Discussion

The author calls the fortis/lenis variants allophones of each other, suggesting the alternating sounds represent the same phonemes. The fact that the relationship between the lenis and fortis members is not elaborated on usually implies it is a minor phonetic change that does not result in a sound that could be confused with other phonemes.[7] Therefore there is ample reason to classify the lenis/fortis variations as not neutralizing.

3.8. Arbore

Language:	Arbore	*Dialect*:	Not specified
Phylum:	Afro-Asiatic, Cushitic, East, Western Omo-Tana		
Inventory:	Kirchner	*Source*:	Harris 1990
Add. Source:	Hayward 1984 (language source for Harris 1990)		
Lenition(s):	Optional debuccalization of coda ejectives and implosives		
Type:	Debuccalization		
Classification:	Neutralization		
Difference:	Added restrictions on the context where this process applies		

[7] Speakers of the language have confirmed that what little variation there is, it is very minor and does affect the sounds enough to be perceived as different phones.

Consonant Inventory

Table 26: Phonemic Consonant Inventory of Arbore (Hayward 1984)

							ʔ
p^{h}		t^{h}		$tʃ^{h}$		k^{h}	
		t'		tʃ'		k'	
b		d		dʒ		g	
ɓ̥		ɗ					
	f		s	ʃ			h
			z				
m		n			ɲ	ŋ	
		ɾ					
		l					
w					j		

Description

Ejective and implosive stops are optionally reduced to glottal stops before a non-identical consonant. In the case of ɓ̥ it only occurs obligatorily when there is a preceding m.

(3.8) Examples of the reduction of ejectives and implosives (Hayward 1984: 64):

a.	/beek'-t-aw/	bee[ʔt̪^{h}]aw / be[ʔet̪^{h}]aw	'my wound'
b.	/gaaɗ-n-e/	ʔina gaa[ʔn]e / ga[ʔan]e	'we buried'

The data in (3.8) also shows the optional "post pharyngeal metathesis" rule that transposes the second vowel mora of a long vowel and an immediately following laryngeal when the latter is followed by an obstruent or nasal.

In addition to the above, the implosives ɗ and ɓ̥ are optionally reduced to glottal stops word-finally, when preceded by a non-syllabic sonorant.

(3.9) Examples (Hayward 1984: 65):

a.	(/bonɓ̥/)	bo[mʔ]	'pond (abs)'
b.	(/saagalɗ/)	saaga[lʔ]	'nine (abs)'

Discussion

The two related processes described above affect the ejectives (t', tʃ', k') and implosives (ɓ̥, ɗ) in Arbore. The phonemic inventory suggests that the ejectives and implosives all contrast with each other, and therefore, when all are reduced to a glottal stop in the same environment the contrast is obliterated. Note, however, that this debuccalization is optional.

Notes on Difference from Kirchner's list

Kirchner (1998) lists the lenition process for Arbore as "debuccalization of coda ejectives", giving the impression that this process is more general and affects more sounds in the language than it actually does. According to Hayward (1985), the language source for Harris (1990), not only are there restrictions on where the debuccalization of coda ejectives can apply, but the process itself is said to be optional. Hayward also includes implosives in his description of the debuccalization rule, which I add.

3.9. Assamese

Language:	Assamese *Dialect*: Eastern and Western
Phylum:	Indo-European, Indo-Iranian, Indo-Aryan, Eastern zone, Bengali-Assamese
Inventory:	Kirchner *Source*: Goswami 1966
Lenition(s):	1. Labial fricatives restricted to _# position 2. p^h t^h $b^{ɦ}$ - ϕ θ β
Type:	1. Distributional observation 2. Spirantization
Classification:	1. Not Neutralization 2. Not Neutralization

Consonant Inventory

Table 27: Phonemic Consonant Inventory of Assamese

p	t			k	
p^h	t^h			k^h	
b	d			g	
$b^{ɦ}$	$d^{ɦ}$			$g^{ɦ}$	
		s		x	
		z			ɦ
m	n			ŋ	
	r				
	l				
w			j		

Description

The first lenition process Kirchner lists for this language is that labial fricatives are restricted to word-final position. ɸ is said to occur finally before any juncture, but in very few words of which most are loans. After /a ɛ/ it freely varies with the affricate pɸ (3.10). The same is true for β, including the variation with bβ, but most words that end with the voiced bilabial are loans from Sanskrit.

(3.10) Examples of ɸ ~ pɸ (p.20):

a.	/ɦaph/	[ɦaɸ] ~ [ɦapɸ]	'half'
b.	/alɛph/	[alɛɸ] ~ [alɛpɸ]	'inverted comma'

The second process is that, in some dialects of Assamese, in final position p^h t^h $b^ɦ$ tend to spirantize to ɸ θ β.

(3.11) p^h - ɸ (p.20):

a.	/kɒph/	[kɒɸ]	'phlegm'
b.	/bɒrɒph/	[bɒrɒɸ]	'ice, snow'

(3.12) $b^ɦ$ - β (p.23):

a.	/lobɦ/	[loβ] ~ [lobβ]	'greed'
b.	/khjobɦ/	[khjoβ] ~ [khjobβ]	'anger'

(3.13) t^h ~ θ (p.30):

a.	/pith/	[pith] ~ [pitθ]	'place, region'

Discussion

The sounds ɸ and β are rare in Assamese, occurring mostly in loan words, and are also restricted to word-final position. As is clear from the second process for this language, these are results of the word-final spirantization of p^h and $b^ɦ$. As with any other static distributional property, the restriction on contexts of certain sounds cannot in itself lead to neutralization.

The final spirantization of p^h t^h $b^ɦ$ to ɸ θ β is also not neutralizing because the fricatives surface in the language only as results of this process and are not alternants of any other phonemes in this position.

Additional Comments

The aspirated voiceless velar is also said to alternate with a fricative in some caste dialects, but this is rare, and in his section on "allophones, their phonetic values, and their distributions" Goswami indicates that, in general, there is no variation for k^h (3.14).

(3.14) k^h (p.39):

a.	/sik^h/	[sik^h]	'the Sikhs'
b.	/dɛk^h/	[dɛk^h]	'to see'
c.	/xuk^h/	[xuk^h]	'happiness'

3.10. Babine

Language:	Babine *Dialect*: Not specified
Phylum:	Na-Dene, Nuclear Na-Dene, Athapaskan-Eyak, Athapaskan, Canadian, Carrier-Chilcotin, Babine-Carrier
Inventory:	Lavoie *Source*: Story 1984
Lenition(s):	1. s → z stem-finally (optional) 2. x → h stem-finally
Type:	1. Voicing 2. Debuccalization
Classification:	1. Not Neutralization 2. Not Neutralization
Trends:	1. Phonetic, but not phonological, neutralization 2. Possible case of phonemic overlap (x → h, h → Ø)
Difference:	Split processes into two and note that the first one is optional

Consonant Inventory

Table 28: Consonant Inventory of Babine

	t	tl̥	ts	(tʃ)	k	q	ʔ
	t^ʔ	tl̥^{ʔ}	ts^ʔ	(tʃ^ʔ)	k^ʔ	q^ʔ	
						q^w	
						q^{wʔ}	
b	d	dl	dz	(dʒ)	g	ɢ	
						ɢ^w	
			s		x	χ	h
						χ^w	
			z		ɣ	ʁ	
						ʁ^w	
	l̥						
	l						
m	n						

The sounds in parentheses (tʃ tʃˀ dʒ) are the Lake Babine phonemes corresponding to the River Babine k kˀ g.

Description

Story describes the first process, the voicing of s, as affecting only a few cases of the possessed noun stems, where stem-finally there is either no change or the s may become z. The second process affects the phoneme x, which is reported to be marginal to the system and to occur only stem finally where it alternates with h.

Discussion

The sound inventory provided in the source does not appear to be phonemic so it is not clear whether the voicing of sibilants is contrastive, but it is safe to assume that it probably is. With just this in mind the stem-final s → z constitutes phonetic neutralization. However, this process is limited to a few cases of the possessed noun stems, and is optional at that, so phonological neutralization is extremely unlikely. In general, when a process affects very few forms in the language it does not obliterate meaning distinctions. The fact that the process is optional makes the chances of obliterating meaning distinctions even smaller.

In Babine x is attested only stem-finally, where the x - h alternation takes place. Neutralization with existing h's in the language would be possible if existing h's could be found stem-finally. However, Story reports that in stem-final position h is lost. Therefore, all surface cases of stem-final h must be the results of debuccalized x, and neutralization with existing sounds is avoided.

Additional Comments

The second process constitutes a type of phonemic overlap, where x → h and h → Ø in the same context.

Notes on Difference from Kirchner's list

Lavoie (1996), cited in Kirchner (1998), lists both of the above processes as one. However, the only thing the two processes have in common is the fact that they apply stem-finally, which is why I split them into two processes. In addition to this, Lavoie's grammar source (Story 1984) reports the first process is optional, a detail I add.

3.11. Badimaya

Language:	Badimaya	*Dialect*:	Not specified
Phylum:	Australian, Pama-Nyungan, South-West, Wadjari		
Inventory:	Kirchner	*Source*:	Dunn 1988
Lenition(s):	1. Stop voicing word-medially 2. Velars rarely voice 3. t̪ c → ð ʒ / V_V		
Type:	1. Voicing 2. Distributional observation 3. Spirantization		
Classification:	1. Not Neutralization 2. Not Neutralization 3. Not Neutralization		
Difference:	The organization of the processes and their descriptions		

Consonant Inventory

Table 29: Phonemic Consonant Inventory of Badimaya

p	t̪	t	ʈ	c	k
m	n̪	n	ɳ	ɲ	ŋ
			ɽ		
		r			
	l̪	l	ɭ	ʎ	
w				j	

Description

An explanation of the sounds Dunn considers phonemic must precede the description of the relevant lenition processes in Badimaya. There is only one set of stops in the language, listed here as p t̪ t ʈ c k; in his inventory, however, Dunn selects the voiced counterparts as the basic phonemes (i.e., b d̪ d ɖ ɟ g). Traditionally, linguists agree that if a language includes only one set of stops it would be the voiceless set.[8] The fact that Dunn selected the voiced, rather than voiceless, stops as phonemic is merely a matter of preference. The significance of including only one of the two sets of stops, illustrating that voicing is not contrastive in the language, is far more important than the arbitrary labeling of

---

[8] According to IMPLICATIONAL HIERARCHY if a language has voiced stops it has voiceless ones (Westbury and Keating 1986); thus an inventory that includes voiced but not voiceless stops is considered unnatural.

one or the other as phonemic. With this in mind, I follow Kirchner in reanalyzing the lenition processes in Badimaya as voicing of the voiceless phonemes, rather than Dunn's version of devoicing of the voiced ones.

The first process in Badimaya involves the word-medial voicing of stops. Dunn reports that word final stops are rare, but when they do occur they are voiceless and sometimes aspirated, and word-initial stops are also usually voiceless (3.15). Word-medially the stops are voiced. Velars are reported to rarely appear as their voiced counterparts, which comprises the distributional observation listed as lenition process 2.

(3.15) Examples of word-initial and word-final stops (p.33):

a.	/t̪alkat/	[t̪algatʰ]	'native rat'
b.	/ŋuʎak/	[ŋʊʎak]	'white-tailed black cockatoo'
c.	/kurara/	[kʊrarʌ]	'kurara tree'

While he does not provide any examples, Dunn also reports that dental and palatal stops (t̪ c) are frequently realized as voiced fricatives (ð ʒ) intervocalically; this is process 3.

Discussion

The phonemic inventory of consonants for Badimaya illustrates that neither spirantization, aspiration, or voicing are contrastive. This means that changes in voicing (e.g. t̪ to d̪), aspiration (e.g. t to tʰ) or spirantization (e.g. d̪ to ð) do not result in sounds that already exist in the inventory because all instances of these sounds in Badimaya occur as results of the lenitions in question. Therefore no contrasts are neutralized by the voicing or spirantization processes. This conclusion remains the same regardless whether one considers the Badimaya phonemes as voiceless (which are voiced in certain environments) or voiced (and are devoiced at word-boundaries).

The fact that velars tend to remain voiceless is a static distributional property of the language that may have resulted from an active lenition process at some point, but is at the time of Dunn's research merely an observation. Since this observation does not constitute an active process in the language it is not capable of neutralizing any contrasts.

Notes on Difference from Kirchner's list

Kirchner (1998) lists 4 lenitions for Badimaya: (1) Stop voicing except _#; (2) In certain words #_ stops have a greater tendency to be voiceless; (3) Greater tendency of velar stops to devoice; and (4) d ɟ → ð ʒ / V_V. I consider the voicing of medial stops and devoicing of stops at word-boundaries as part of the same process (process 1).

3.12. Balti (Purki)

Language:	Balti	*Dialect*:	Purki
Phylum:	Sino-Tibetan, Tibeto-Burman, Himalayish, Tibeto-Kanauri, Tibetic, Tibetan, Western		
Inventory:	Kirchner	*Source*:	Rangan 1975
Add. Source:	Read 1934		
Lenition(s):	d̪ → ð / V_V; ð ʁ f are restricted in where they can appear		
Type:	Spirantization (& Distributional observation)		
Classification:	Not Neutralization		
Difference:	Process description is corrected		

Consonant Inventory

Table 30: Phonemic Consonant Inventory of Purki Balti (Rangan 1975)

p		t̪	t	ts		cç	k	q	
p^h		t̪^{h}	t^h	ts^h		cç^h	k^h		
b		d̪	d			ɟʝ	g		
	(f)			s	ç			χ	h
				z	ʝ			ʁ	
m			n		ɲ		ŋ		
			ɾ						
			l						
w					j				

Rangan never explains the parentheses around f but Read reports that f is used in foreign and borrowed words only and is therefore not a native phoneme in the language.

Description

Rangan reports that ð is an allophone of d̪ which occurs intervocalically. He does not provide examples to show that there is an active synchronic alternation, but he does show that ð surfaces in intervocalic position.

(3.16) Intervocalic ð (Rangan 1975: 84):

a. [baːðem] 'almond'

b. [aðit̪] 'Sunday'

The fricatives ð ʁ f are restricted to certain positions in words: ð only occurs medially, ʁ occurs initially and medially, and f occurs only non finally. Word-finally, of all the aspirates, fricatives and affricates, only ç, χ and cç can occur (Rangan 1975).

Discussion

The fact that ð is restricted to medial position appears to be a result of the intervocalic spirantization of d̪. As I mention above, Rangan does not provide examples to show that the spirantization of d̪ is active, but he does appear to be confident that the spirantization did occur. This particular process does not obliterate any contrasts in the language because ð is not a phoneme in Balti so there are no existing ð's with which the spirantized d̪ could neutralize.

Judging by the behavior of the dental sound, and the similar restriction on contexts of the other two fricatives (f ʁ) it is possible that they were also the result of intervocalic spirantization at some point in the language's history. It is also possible, based on the fact that f is reported to be a non-native phoneme, that it is restricted in position for other reasons than spirantization of p. This is not a question that can be answered in the present analysis, but it is also not necessary to know the origin of the restriction on spirants in the language in order to determine the functional classification of this phenomenon. The fact is, a static distributional observation such as this cannot obliterate contrast because it is not an active process and can never result in sounds that may be confused with existing phonemes in the language.

Notes on Difference from Kirchner's list

Kirchner (1998) lists the lenitions in Balti as three separate processes, one for the distributional property of each fricative (ð ɣ f). I combine the three observations, correcting the velar fricative to the uvular one as reported by Kirchner's grammar source (Rangan 1975). I also add to this process a relevant detail, that ð results from the spirantization of d̪.

3.13. Bashkir

Language:	Bashkir	*Dialect*:	Not specified
Phylum:	Altaic, Turkic, Western, Uralian		
Inventory:	Lavoie	*Source*:	Poppe 1964
Lenition(s):	b → β in fast speech		

Type:	Spirantization
Classification:	Not Neutralization
Difference:	The context of this process is corrected

Consonant Inventory

Table 31: Consonant Inventory of Bashkir

				(tʃ)	(cç)			(ʔ)
p^h		t^h				k^h	q^h	
b		d				g		
	(f)	θ	s		ç	x	χ	h
	(v)	ð	(z)		(ʝ)		ʁ	
m		n				ŋ	ɴ	
		r						
		l ɫ						
w					j			

The sounds in parentheses occur only in loan words (most of which come from Russian or Arabic). Additional details about the inventory include that ɫ is in complementary distribution with l occurring in the vicinity of back vowels, χ ɴ are also both optional allophones of their respective velar sounds (x ŋ) in the vicinity of back vowels.

Description

Poppe reports that the articulation of the unaspirated bilabial stop b is weak due to the fact that the lips do not produce a complete closure which, in fast speech, results in β. In addition to this, the b does not occur in syllable or word final positions.

Discussion

The source is very short on details, but it is clear that the spirantization of b does not neutralize any contrasts in the language because the bilabial fricative β surfaces in Bashkir only as a result of this alternation. The closest existing sound in the language is v, which although similar to β, is made at a different place of articulation. Moreover, v is not a native phoneme in Bashkir.

Notes on Difference from Kirchner's list

Citing Lavoie (1996) Kirchner (1998) lists the lenition process in Bashkir as b → β / V_V. However, I follow Lavoie's language source (Poppe 1964), who reports that the process occurs in fast speech and mentions nothing about it being limited to intervocalic context.

3.14. Basque (Souletin)

Language: Basque — *Dialect*: Souletin
Phylum: Basque
Inventory: Lavoie — *Source*: Hualde 1993
Lenition(s): Spirantize voiced stops
Type: Spirantization
Classification: Not Neutralization
Difference: One process omitted; description of 2nd process is corrected

Consonant Inventory

Table 32: Phonemic Consonant Inventory of Souletin Basque

p		t	ts	tʃ	c	cç	k		
p^h		t^h					k^h		
b		d		(dʒ)	ɟ		g		
	f		s	ʃ	ç			h	ħ
			z	ʒ	ʝ				
m		n			ɲ				
		r							
		l			ʎ				

The sound in parenthesis (dʒ) appears only in borrowings; z and ʒ also mostly appear in borrowed vocabulary.

Description

The voiced stops b d g are realized as continuants in "the same context as standard Spanish" (p.298), which is described as everywhere except after a pause, a nasal, and in the case of d also after a lateral. The voiced palatal stop ɟ can also spirantize, but more rarely.

Discussion

There are three distinctive series of stops in Souletin Basque: the voiceless unaspirated (p t c k), the voiceless aspirated (p^h t^h k^h), and the voiced (b d ɟ g). The opposition is manifested only in voicing and aspiration, so spirantization is not contrastive in the language. Hence, no meaning distinctions are obliterated when voiced stops spirantize in certain, or rather most, environments.

Notes on Difference from Kirchner's list

Lavoie (1996), cited in Kirchner (1998), lists two lenition processes in Basque: k → ɣ word-finally, and spirantization of voiced fricatives intervocalically. Lavoie's grammar source (Hualde 1993) lists neither the first nor the second as processes in the language. For this reason the first process is omitted from my list, and the second is rewritten as the spirantization of voiced stops in Souletin Basque, a process that is described in the source (and was probably the one Lavoie intended to list, obscured by a typo).

3.15. Blackfoot

Language:	Blackfoot	*Dialect*:	Not specified
Phylum:	Algic, Algonquian, Plains		
Inventory:	Lavoie	*Source*:	Frantz 1971
Lenition(s):	w j - Ø		
Type:	Loss		
Classification:	Not Neutralization		
Trends:	Phonetic but not phonological neutralization (of glides with Ø)		
Difference:	Omitted processes from an unavailable source		

Consonant Inventory

Table 33: Phonemic Consonant Inventory of Blackfoot

p	t			k	?
		s		x	
m	n				
w			j		

Description

The glides are lost in certain environments in the language, as is formalized by Frantz in the following rule (but no examples are provided):

(3.17) Glide loss (p.8):
$\{w, j\} \rightarrow \emptyset / \{C, \#\}\_$ [condition: $C \neq ʔ$]

Discussion

The loss of glides appears to manifest itself only in the suffixes of some possessives. E.g., in "obligatorily possessed nouns" third person suffix -wa, for forms possessed by 1 or 2, is reduced to -a (/C_); and the final -ji is reduced to -i (/C_) in the forms with 3 as possessor (p.23).

The loss of the glides w and j in similar environments does lead to the phonetic neutralization between w j and Ø. However, since this process is limited to two suffixes, and is additionally restricted by the preceding sounds, phonological neutralization is surely avoided because no obliteration of meaning distinctions is possible by a process that affects so few forms.

Notes on Difference from Kirchner's list

Lavoie (1996), reprinted in Kirchner (1998), lists 4 lenition process for Blackfoot, based on research from two sources for the language (Frantz 1971 and Proulx 1989). The first process is listed here and is based on information in Frantz. The other 3 processes were based on research by Proulx which was not available to me; since these processes could not be verified or researched in their primary source they are omitted.

3.16. Bontoc

Language:	Bontoc — *Dialect*: Not specified
Phylum:	Austronesian, Malayo-Polynesian, Western Malayo-Polynesian, Northern Philippine, Northern Luzon, South-Central Cordilleran, Central Cordilleran, Nuclear Cordilleran, Bontok-Kankanay
Inventory:	Lavoie — *Source*: Reid 1971
Add. Source:	Sadenadel 1909
Lenition(s):	b d g → f ts k^h syllable-initially
Type:	Devoicing (& Spirantization/Aspiration)
Classification:	Not Neutralization

Consonant Inventory

Table 34: Consonant Inventory of Bontoc (Reid 1971)

p	t			k	q
b	d			g	
		s			
m	n			ŋ	
	l				
w			j		

Reid also reports that r is an allophone of l, and u: is an allophone of j.

Description

According to Reid, f ts k^h are the syllable initial allophones of b d g. Sadenadel (1909), an additional source I consult, suggests that p, b and f are allophones of the same phoneme, and that similar variations of t and k also exist.

Discussion

The information provided by the primary source for the language (Reid 1971) is limited to identifying the voiced stops b d g and their devoiced (and spirantized/aspirated) counterparts as allophones of the same phoneme. Spirantization (or, in the case of the velar, aspiration) are therefore not contrastive in the language, so the lenition of b d g is not neutralizing.

According to Sadenadel (1909) p b f are all one phoneme, as are t d ts, and k g k^h; this would suggest that neither voicing nor spirantization is contrastive. Regardless of whether this is true or not, this source does not contradict the finding that the process listed for Bontoc (devoicing and spirantization) does not neutralize existing contrasts in the language.

3.17. Bulgarian

Language: Bulgarian *Dialect*: Contemporary Standard
Phylum: Indo-European, Slavic, South, Eastern
Inventory: Kirchner *Source*: Scatton 1984
Lenition(s): 1. vN → ṽ / _ C[+cont]
2. Voicing assimilation in medial clusters
3. Devoicing at end of phonological words
4. Degemination across morpheme boundary (optional)

5. ts x → dz ɣ / _ -son, +voi

Type: 1. Nasalization & Loss
2. Voicing assimilation
3. Devoicing
4. Degemination
5. Voicing

Classification: 1. Not Neutralization
2. Neutralization
3. Neutralization
4. Not Neutralization
5. Not Neutralization

Trends: Process 1 involves loss that is not neutralizing

Difference: A few of the contexts are corrected

Consonant Inventory

Table 35: Consonant Inventory of Bulgarian

p	pʲ			t	tʲ	ts	tsʲ	tʃ	c	k
b	bʲ			d	dʲ			dʒ	ɟ	
		f	fʲ			s	sʲ	ʃ		x
		v	vʲ			z	zʲ	ʒ		
m	mʲ			n	nʲ					
				r	rʲ					
				l	lʲ					
									j	

Description

Before a nasal consonant that is followed by a fricative, a vowel is nasalized (3.18a), and the nasal consonant may be lost (3.18b).

(3.18) Nasalization of a vowel & loss of nasal (p.57):

a.	/onzi/	[õnzi]	'that'
b.		[õzi]	'that'

Process 2 is the assimilation in voicing of obstruents to following obstruents (except when they are v vʲ, neither of which trigger assimilation). Very similar to the 2nd process, process 3 is final devoicing, where word-final obstruents are always voiceless (3.19).

(3.19) Final Devoicing (p.73):

grat 'city' grad-ove 'cities'

Process 4 is optional degemination across morpheme boundaries. In native lexical items there is a strong tendency to reduce geminates to singletons, and in foreign borrowings, where geminates are rare the tendency to degeminate is very great.

Process 5 is listed in the source as "/x/ and /ts/ have voiced allophones when they stand (rarely) before voiced obstruents" (p.18).

Discussion

Process 1 involves the loss of a nasal sound between a vowel and a fricative with the subsequent nasalization of the vowel. Nasalization in vowels is not distinctive in Bulgarian, so no contrast is obliterated. That is, the resulting nasal vowels do not neutralize with existing vowels in the language. The loss of the nasal sound also does not lead to neutralization. The distinction between the existence of a nasal and Ø is maintained in the nasalized vowel, so phonological neutralization does not occur. This is yet another example in the corpus of a phonetic neutralization (between Ø and the lost n which results in Ø) that does not lead to the obliteration of meaning distinctions.

Process 2 involves the assimilation in voicing of medial clusters. Since voicing in Bulgarian obstruents is distinctive, both voicing and devoicing of sounds, as is the case with assimilation, leads to neutralization of the voicing opposition. This process is very common in Slavic languages and it affects the coda. Similarly process 3, the devoicing of final obstruents, neutralizes the distinction between voiced and voiceless sounds.

Process 4, degemination across morpheme boundary, which incidentally is the only context where gemination occurs in native vocabulary, is not neutralizing because gemination is not contrastive in Bulgarian. The description of degemination in the source appears to be a reading, rather than phonological, rule: a learner of Bulgarian is told to ignore geminates in the spelling and pronounce them as singletons.

Process 5, the voicing of ts and x before voiced obstruents is rare, and it is allophonic. The voiced variants of these two sounds (dz and ɣ) are not included in the inventory of Bulgarian sounds, which suggests the only occurrences of these sounds are as results of the voicing process in question. This process can be categorized as part of the general voicing assimilation in the language, but since it is the only such assimilation that does not lead to neutralization I follow both Scatton and Kirchner in listing it separately.

Notes on Difference from Kirchner's list

The contexts for some of the lenition processes for Bulgarian are rewritten from the way they appear in Kirchner (1998) to more closely reflect the description in Kirchner's grammar source (Scatton 1984).

3.18. Canela-Krahô

Language:	Canela-Krahô	*Dialect*: Not specified
Phylum:	Macro-Ge, Ge-Kaingang, Ge, Northwest, Timbira	
Inventory:	Lavoie	*Source*: Popjes & Popjes 1986
Lenition(s):	1. l → ɾ / V_V, #_, C_ 2. p t k → b d g in syllable edges 3. j x → h phrase initially	
Type:	1. Flapping 2. Voicing 3. Debuccalization	
Classification:	1. Not Neutralization 2. Not Neutralization 3. Not Neutralization	
Trends:	Phonetic but not phonological neutralization (in process 3 between j and x)	
Difference:	Contexts for all the processes are corrected	

Consonant Inventory

Table 36: Phonemic Consonant Inventory of Canela-Krahô

p	t	ts		k	
				k^h	
					h
m	n			ŋ	
	l				
w			j		

Description

In Canela-Krahô l has a flap allophone which occurs intervocalically, utterance-initially and following consonants (Process 1). The voiceless unaspirated stops p t k are all voiced syllable initially in unstressed syllables following voicing, and syllable finally preceding voicing (Process 2). Also, phrase-medial j and x surface as h phrase initially (3.20) (Process 3).

(3.20) Morpho-phonological alternation j x - h (p.193):

a.	pɔ xũmrɛ	'male deer'	hũmrɛ	'male, man'
b.	pjen xom	'grains of sand'	hom	'kernels'
c.	ijõ rɔp	'my dog'	hõ rɔp	'his dog'
d.	ijɑpen	'I worked'	hɑpen	'he worked'

Discussion

Process 1, the alternation between the lateral and a flap, does not obliterate any contrasts in the language because the flap only occurs in Canela-Krahô as a result of this process and therefore does not neutralize with existing sounds. Process 2, the syllable initial and final voicing of simple stops is also not neutralizing because voiced stops only occur as a result of this process and therefore cannot be confused with existing sounds in the language.

Process 3, the alternation between j x and h, requires some elaboration. In particular, the fact that h has two allophones, a front velar fricative x which occurs before high oral vowels, and a glottal stop ʔ which occurs syllable final preceding consonants. Phonetically, the j x - h alternation neutralizes the difference between j and x word-initially. But phonologically there are two details to consider: (1) x is an allophone of h, and (2) all cases of j alternating with h are limited to function words. For this reason there is good reason to believe that the functional load of the j/x (or rather, j/h) opposition is small, as is usually the case when one of the distinctive sounds is limited to function words. Thus, the sound merger between j and h does not lead to phonological neutralization because, as only function words are affected, the consequence to communication is minimal (see §2.1.3.1 for a discussion of this point).

Notes on Difference from Kirchner's list

Lavoie (1996) describes the first process as l → ɾ medially whereas I adhere to her grammar source (Popjes & Popjes 1986), which describes the lateral-flap alternation as occurring intervocalically, utterance-initially, and following consonants. The second process is described as "stops voiced in medial position", and again, I follow the language source which provides a more complex context for this process. Finally, the third process is described as occurring initially, which normally implies "word-initially", but the authors specify that the j x - h alternation is limited to phrase- (not word-) initial position.

3.19. Car Nicobarese

Language: Car Nicobarese *Dialect*: Not specified
Phylum: Austro-Asiatic, Mon-Khmer, Nicobar, Car
Inventory: Kirchner *Source*: Das 1977
Lenition(s): r → d / n, l _
Type: Occlusivization
Classification: Not Neutralization

Consonant Inventory

Table 37: Consonant Inventory of Car Nicobarese

p		t		ʈ	tç	k	
		d					
	f		s				h
m		n			ɲ	ŋ	
		ɾ					
		r					
		l					

Das notes that the retroflex ʈ occurs as an allophone of the alveolar voiceless stop, but does not provide the context where the alternation takes place.

Description

In Car Nicobarese the d occurs only as an allophone of r when it is preceded by n or l.

Discussion

The source is very short on details, but the r - d alternation is specifically listed as allophonic. The d is reported to only surface as a result of this alternation so neutralization with existing sounds in the language is not possible.

3.20. Catalan (Eastern)

Language: Catalan *Dialect*: Eastern
Phylum: Indo-European, Italic, Romance, Italo-Western, Western, Gallo-Iberian, Ibero-Romance, East Iberian

Inventory: Lavoie *Source*: Hualde 1992
Lenition(s): b d g → β ð ɣ medially
Type: Spirantization
Classification: Not Neutralization

Consonant Inventory

Table 38: Phonemic Consonant Inventory of Eastern Catalan

p		t	ts	tʃ		k
b		d	dz	dʒ		g
	f		s	ʃ		
			z	ʒ		
m		n			ɲ	
		ɾ/r				
		l			ʎ	

Description

Hualde writes that

> /b d g/ have noncontinuant [b d g] and continuant [β ð ɣ] realizations. In a syllable onset the noncontinuant allophones appear after a pause, after a plosive, after a nasal, and in the case of /d/ after a lateral. /b/ is also noncontinuant after /f/. Elsewhere, in a syllable onset, these segments are realized as continuant. In coda position, the noncontinuant allophones appear. (p.368)

Discussion

Hualde describes the Catalan approximants as allophones of the voiced stops, and does not include the approximants in the list of phonemes in the language. This makes it clear that β ð ɣ only surface in the language as a result of the spirantization process described here, so neutralization with existing sounds is not possible.

3.21. Cuna

Language:	Cuna[9]	*Dialect*:	Not specified
Phylum:	Chibchan, Kuna		
Inventory:	Kirchner	*Source*:	Keating et al. 1983
Lenition(s):	p t k → b d g (blocked in geminates)		
Type:	Voicing		
Classification:	Not Neutralization		

Consonant Inventory

Not provided in the source; Keating et al.'s language source (Holmer N, 1947, "Critical and Comparative Grammar of the Cuna Language") was not available.

Description

Keating et al. write that

> [I]n Cuna the initial voiceless unaspirated [p t k] are not part of a [voice] contrast, but in medial position they become voiced and contrast with voiceless geminates [pː tː kː]. (p.279)

Discussion

In Cuna there is a contrast between singletons and geminates. The voicing of singletons in medial position does not lead to neutralization with the geminates (which are neither voiced nor degeminated in medial position).

(3.21) Positional allophones in Cuna (p.286):

phonemes	initial allophones	medial allophones	final allophones
/p/ - /pp/	[p] [b̥]	[b] - [pː]	[p˚]
/t/ - /tt/	[t] [d̥]	[d] - [tː]	[t˚]
/k/ - /kk/	[k] [g]	[g] - [kː]	[k˚]

According to this study there is no voicing contrast in Cuna so the voiced singletons do not neutralize with existing sounds (i.e., the only medial b d g in the language are the voiced p t k).

---

9 Most likely the Cuna language discussed here and Kuna, whose dialect Paya is discussed in §3.59, refer to the same language.

3.22. Dahalo

Language:	Dahalo	*Dialect*:	Not specified
Phylum:	Afro-Asiatic, Cushitic, South		
Inventory:	Lavoie	*Source*:	Tosco 1991
Lenition(s):	b d → β ð / V_V		
Type:	Spirantization		
Classification:	Not Neutralization		

Consonant Inventory

Table 39: Phonemic Consonant Inventory of Dahalo

p		t̪	t		ts		cç	k	k^w		ʔ
p'			t'	tl'	ts'		cç'	k'	k^{w}'		
b		d̪	d	dl	dz		ɟʝ	g	g^w		
							ɟʝ'				
^{m}b		^{n}d̪	^{n}d		^{n}dz		^{ɲ}ɟʝ	^{ŋ}g	$^ŋg^w$		
ɓ		ǀ	ɗ								
		^{n}ǀ									
	f				s	ʃ				ħ	h
	v									ʕ	
m			n				ɲ				
			r								
			l								
			ɬ								
w							j				

Description

In Dahalo intervocalic b and d are spirantized to β and ð.

Discussion

Tosco lists β and ð as non-phonemic in the language, and based on this the result of the spirantization of b and d is not neutralizing because it does not obliterate any meaning distinctions in Dahalo.

3.23. Djabugay

Language:	Djabugay	*Dialect*:	Not specified

Phylum: Australian, Pama-Nyungan, Yidinic
Inventory: Kirchner *Source*: Patz 1991
Lenition(s): 1. r → ɾ / [stressed]V_V
2. p t c k → b d ɟ g / #_
Type: 1. Flapping
2. Voicing
Classification: 1. Not Neutralization
2. Not Neutralization
Difference: Descriptions of processes are corrected

Consonant Inventory

Table 40: Phonemic Consonant Inventory of Djabugay

p	t		c	k
m	n		ɲ	ŋ
		ɽ		
	r			
	l			
(w)			j	w

Description

Process 1 is the tapping of the trill: "In normal speech the trill is often reduced to a tap [ɾ] between a stressed and an unstressed vowel" (p.253). This is most obvious between two low vowels (3.22a) less so between identical high vowels (3.22b), and least strong between different vowels (3.22c) where the trill and the tap appear to be in free variation in normal speech.

(3.22) r → ɾ / V_V (p.253):

a.	garaj	['gaɾaj]	'to come'
b.	biri		'near'
c.	dira		'teeth'

Process 2 is the word-initial voicing of p t c k to b d ɟ g. This process is actually described very differently in the source because the author considers the voiced stops to be the basic phonemes that are devoiced in most environments. I follow a more conventional route by describing the basic phonemes as the voiceless series that are voiced in a limited environment, which appears to be restricted to word-initial position.

(3.23) An example of word-initial voicing (p.253):

pulpal [bulpʌl] 'to rub'

Discussion

Process 1, the alternation between the trill and the tap in Djabugay is a straightforward case of allophony where no contrast is neutralized because all cases of intervocalic taps in the language are results of this lenition process.

Process 2, the voicing of simple stops in Djabugay also does not neutralize any contrasts in the language. The inventory of phonemes makes it clear that voicing is not contrastive for stops, and therefore, all cases of voiced stops in the language must have resulted from this voicing process and do not neutralize with existing sounds.

Notes on Difference from Kirchner's list

Kirchner (1998) lists the lenition processes for this language as d - ɾ and r - ɾ in the same environments. The fact that both sounds alternate with ɾ suggests a case of neutralization. The tapping of d, however, is not found in Kirchner's grammar source (Patz 1991) and is therefore omitted. Patz does discuss d - t which is included and rephrased as initial voicing, as discussed above. The context for the flapping of r is also corrected according to the description in Patz.

3.24. Efik (Calabar-Creek)

Language:	Efik *Dialect*: Calabar-Creek Town
Phylum:	Niger-Congo, Atlantic-Congo, Volta-Congo, Benue-Congo, Cross River, Delta Cross, Lower Cross, Obolo, Efik
Inventory:	Lavoie *Source*: Dunstan 1969
Lenition(s):	b d k → β ɾ ɣ non-initial in a word stem and before a vowel
Type:	Spirantization
Classification:	Not Neutralization
Trends:	Flapping of alveolar patterns with spirantization

Consonant Inventory

Table 41: Consonant Inventory of Calabar-Creek Efik

		t			k	kp
b		d				
	f		s			
m		n		ɲ	ŋ	
w				j		

In his inventory of sounds Dunstan also includes p as a pre-pausal allophone of b, g and ɣ as allophones of k, ɾ as an allophone of d, and $ŋ^w$ as an allophone of ŋ.

Description

In Efik b d k have various allophones, depending on their position in the word. The positions that matter are illustrated below:

(3.24) Initial in the word or initial in the stem following a prefix (p.38):

a.	/b/ - [b]	i˩ba˩	'two'
b.	/d/ - [d]	n˧da˧m	'raffia rope'
c.	/k/ - [k]	ka˧ŋ	'deny!'

(3.25) Non-initial in a word stem and before pause (p.38):

a.	/b/ - [p̥]	ka˧p̥	'carve/dig out!'
b.	/d/ - [t̥]	be̥˧t̥	'wait!'
c.	/k/ - [k̥]	wa˧k̥	'abound/multiply!'

(3.26) Non-initial in a word stem and before a consonant (p.38):

a.	/b/ - [p]	n˧de̥˧p-ke̥˧	'I am not buying'
b.	/d/ - [t]	n˧do˩t-ke̥˧	'I am not worthy'
c.	/k/ - [k]	mŋ˧kpo˧k-ke̥˧	'I am not carving (meat)'

(3.27) Non-initial in a word stem and before a vowel (p.38):

a.	/b/ - [β]	dwo˩β-e˩ba˩	'twelve'
b.	/d/ - [ɾ]	i˧ko̥˧ɾ-e̥˩kpe̥˩ne̥˧	'name of a town'
c.	/k/ - [ɣ]	u˧fo̥˩ɣ-u˧dwa˩	'market store'
d.	- [g]	ti˧ga˧	'kick/shoot!'

[g] only occurs after /i/ in non-initial position and only in some dialects.

Discussion

The spirantization of b d k in Efik does not obliterate any contrasts because β ɾ and ɣ only surface in the language as results of this lenition process and therefore do not neutralize with existing sounds.

3.25. English (American)

Language: English *Dialect*: American
Phylum: Indo-European, Germanic, West, English
Inventory: Kirchner *Source*: Kahn 1976
Add. Sources: Chomsky 1964, Joos 1942, Halle 1964a
Lenition(s): Flapping of t, d intervocalically, post-stress
Type: Flapping
Classification: Not Neutralization
Trends: Phonetic, but not phonological, neutralization (t d - ɾ)

Consonant Inventory

Table 42: Consonant Inventory of American English (Halle 1964a)

p		t		tʃ		k	
b		d		dʒ		g	
	f	θ	s	ʃ			h
	v	ð	z	ʒ			
m		n					
		r					
		l					
w					j		

Description

In American English intervocalic t and d both flap (to ɾ) in post-stress position.

Discussion

In discussing the flapping of t and d, Kahn notes that the contrast between the two sounds is maintained in the quality of the vowel:

> Vowels are shorter before voiceless consonants, and in some cases have different quality, and thus may prevent homophony in pairs of words in which

> /t/ and /d/ are pronounced identically. (Cf. Chomsky 1964). For example, although I am convinced that my flapped /t/ and /d/ are identical, *plotter* and *plodder* are not homophonous, the latter word having a distinctly longer vowel, perhaps with slightly different quality. (p.98)

Another example of a change in vowel quality that follows tapping comes from Joos (1942), who reports that in the change from *write* to *writer*, in addition to the change in the intervocalic consonant, there is a shift from [ɐɪ] to [aɪ]. Thus, the phonetic neutralization between t and d, when both are flapped to ɾ in post-stress intervocalic context, does not lead to phonological neutralization because the contrast is shifted to the quantity of the preceding vowel.

3.26. English (British)

Language:	English	*Dialect*:	British
Phylum:	Indo-European, Germanic, West, English		
Inventory:	Lavoie	*Source*:	Milroy, Milroy & Hartley 1994
Lenition(s):	t → ʔ / V_V		
Type:	Debuccalization		
Classification:	Not Neutralization		
Difference:	Environment of first process is corrected; number of processes is different		

Consonant Inventory

Not provided in the source

Description

In the British dialect of English there is intervocalic debuccalization of t, resulting in a glottal stop ʔ.

Discussion

The language source for British English is short on details, but it makes clear that the only glottal stop in the inventory is the one that results from the intervocalic debuccalization of t. Therefore this process does not neutralize any existing contrasts in the language.

Notes on Difference from Kirchner's list

Kirchner (1998) lists 2 lenition processes that Lavoie (1996) attributes to her source for this language (Milroy, Milroy & Hartley 1994): t - ʔ and deletion of stops in final position. I omit the 2nd process from my inventory because it is only mentioned in Milroy et al. as a citation by a different author (Collins & Mees 1990) discussing a different dialect. I also correct the context of the first process according to the grammar.

3.27. English (Cardiff)

Language: English *Dialect*: Cardiff
Phylum: Indo-European, Germanic, West, English
Inventory: Kirchner *Source*: Collins & Mees 1990
Lenition(s): 1. b → β medially
2. b d g dʒ → p t k tʃ / _#
Type: 1. Spirantization
2. Devoicing
Classification: 1. Not Neutralization
2. Not Neutralization
Trends: Process 2 is phonetic, but not phonological, neutralization

Consonant Inventory

Table 43: Consonant Inventory of Cardiff English

p		t		tʃ		k	
b		d		dʒ		g	
	f	θ	s	ʃ		(x)	(h)
	v	ð	z	ʒ			
m		n				ŋ	
		r					
		l					
		(ɫ)					
w					j		

Description

Process 1 is described as a tendency for medial voiced plosives to weaken to fricatives, especially noticeable with b which may be realized as β (3.28). The source is short on details, but notice that the b spirantizes to β and not to v.

(3.28) Weakening of b (p.89):
horrible ['arəβl̩]

Process 2 is the final devoicing of b d g dʒ to p t k tʃ: "/b d g dʒ/ are markedly devoiced in final position and contrast with /p t k tʃ/ is mainly indicated by the length of the preceding vowel" (p.89).

Discussion

Although process 1 is phrased as a general phenomenon that concerns all medial voiced plosives, the only example provided in the source affects the bilabial. The spirantization of the bilabial does not obliterate any meaning distinctions because the β only surfaces as a result of this weakening. It is not clear how the alveolar is affected, but the potential spirantization of the velar would also avoid neutralization because ɣ is not part of any other phoneme in the language.

Process 2 is a case of meaning-maintaining sound merger: although there is phonetic neutralization word finally between the voiced and voiceless series, neutralization is avoided because the contrast is shifted to the length of the preceding vowel.

3.28. English (Cockney)

Language:	English	*Dialect*: Cockney
Phylum:	Indo-European, Germanic, West, English	
Inventory:	Lavoie	*Source*: Andrésen 1968
Lenition(s):	t → ʔ / V_V, _n, _m, _l	
Type:	Debuccalization	
Classification:	Not Neutralization	
Difference:	Context is corrected	

Consonant Inventory

Not provided in the source

Description

In Cockney English simple stops are preglottalized (p t k → ʔp ʔt ʔk) intervocalically and before n, m, l. Andrésen's study suggests that a lot fewer of the intervocalic p and k sounds undergo the process than t's, and that the t will often debuccalize completely (p → ʔ).

Discussion

The glottal stop is not phonemic in English, and therefore the debuccalization of t in certain environments does not neutralize with an existing phoneme in the language. Furthermore, neutralization is avoided with the result of the preglottalization of p and k because these two sounds do not fully lose their oral quality.

Notes on Difference from Kirchner's list

Lavoie (1996), as cited in Kirchner (1998), reports the debuccalization process as applying in intervocalic environment and *after* n, m, l. I follow her grammar source (Andrésen 1968) in describing the process as occurring intervocalically and *before* n, m, l.

3.29. English (Liverpool)

Language:	English	*Dialect*:	Liverpool
Phylum:	Indo-European, Germanic, West, English		
Inventory:	Kirchner	*Source*:	Wells 1982b
Lenition(s):	θ̪ ð̪ → t d / _#, V_V		
Type:	Occlusivization		
Classification:	Not Neutralization		
Difference:	Process is restated		

Consonant Inventory

Not provided in the source

Description

Discussion of this process in the source is limited to the following quote:

> Some [variables] are no doubt to be attributed to Irish influence: the most obvious of these is the use by some speakers of dental or alveolar stops for /θ, ð/, as in [t̪ɾɪi] *three*, [tɹuːt̪] *truth*, [mʊnt̪(θ)] *month*, [d̪at ~ dat] *that*. (p.371)

Wells adds that this process was found to be virtually restricted to working-class Catholics.

Discussion

Except in the case of the very frequent word *that*, where there is a variation between the dental and non-dental d, contrast between existing sounds t and d and the results of the above process (t̪ and d̪) is maintained. Potential phonetic neutralization in more common words like *that* does not lead to loss of meaning distinctions (see discussion of *function words* in §2.1.3.1), and elsewhere phonetic neutralization is avoided, so the process in question does not lead to obliteration of any meaning distinctions.

Notes on Difference from Kirchner's list

In his inventory of lenition processes that fit his definition, Kirchner (1998) lists the process in Liverpool English as t d - θ̪ ð̪, suggesting that there is intervocalic spirantization of stops, a rather common form of lenition. Kirchner's grammar source (Wells 1982b), however, reports the opposite: the Liverpool dialect actually substitutes the stops for the spirants.

3.30. English (London, Leeds & Fife)

Language:	English	*Dialect*:	London, Leeds & Fife
Phylum:	Indo-European, Germanic, West, English		
Inventory:	Kirchner	*Source*:	Harris 1990
Lenition(s):	t → ʔ word-finally		
Type:	Debuccalization		
Classification:	Not Neutralization		

Consonant Inventory

Not provided in the source

Description

In the London, Leeds & Fife dialects of English there is word-final debuccalization of t to a glottal stop ʔ.

Discussion

There is almost no data in the source on this process, but similar processes in other dialects of English discussed here have already established that debuccalization of t is not a neutralizing process because ? only surfaces in English dialects as a result of this process.

3.31. English (New Zealand)

Language:	English	*Dialect*:	New Zealand
Phylum:	Indo-European, Germanic, West, English		
Inventory:	Lavoie	*Source*:	Holmes 1994
Lenition(s):	t → ɾ / V_V		
Type:	Flapping		
Classification:	Not Neutralization		
Trends:	Phonetic, but not phonological, neutralization (t/d)		

Consonant Inventory

Not provided in the source

Description

New Zealand English is exhibiting "the first distinctively American phonetic innovation likely to spread in time to all accents of English (Wells 1982a: 250)", which is t voicing or flapping in intervocalic position.

Discussion

Holmes' work is concerned with describing the extent to which New Zealanders from different social backgrounds flap their t's. While no information is included about the system of contrasts in Holmes' work, this exact process, borrowed from American English, is already familiar to us (see §3.25 on American English). The flapping or voicing of intervocalic t is a non neutralizing process where the contrast between intervocalic t and d is shifted to the preceding vowel.

3.32. Estonian (Southern)

Language:	Estonian	*Dialect*:	Southern
Phylum:	Uralic, Finno-Ugric, Finno-Permic, Finno-Cheremisic, Finno-Mordvinic, Finno-Lappic, Baltic-Finnic		
Inventory:	Kirchner	*Source*:	Harms 1962

Lenition(s):	Non-labiodental obstruents are voiced intervocalically, and in other contexts
Type:	Voicing
Classification:	Not Neutralization

Consonant Inventory

Table 44: Phonemic Consonant Inventory of Southern Estonian

p		t̪	t			k	
	f	s̪		s	ç		h
m		n̪	n			ŋ	
			r				
		l̪	l				
w					j		

Description

Describing allophone distributions in Estonian Harms writes

> All stops and fricatives except /f/ have voiced lenis allophones in the environments (a) V_V, (b) VV_#, (c) R_, (d) _RV. /h/ is not found in environments (b) and (c); /ç/ is not found in (d). (p.17)

Discussion

The voicing of obstruents (except for f) in Estonian is not neutralizing because voicing is not distinctive in the language. All voiced obstruents are results of this lenition process and do not neutralize with existing sounds.

3.33. Faroese

Language:	Faroese	*Dialect*:	Not specified
Phylum:	Indo-European, Germanic, North, West Scandinavian		
Inventory:	Kirchner	*Source*:	Lockwood 1964
Add. Sources:	Strodach 1976, Barnes 2001, Staksberg p.c.		
Lenition(s):	p t c k → b d ɟ g / V_V, _liquids		
Type:	Voicing		
Classification:	Not Neutralizing		
Difference:	One process is omitted		

Consonant Inventory

Table 45: Consonant Inventory of Faroese (Lockwood 1964)

p		t				c	k	
p^h		t^h				c^h	k^h	
	f		s	ʃ	ʂ			h
	v							
m		n				ɲ	ŋ	
		r						
		l				ʎ		
w						j		

Lockwood lists the stop series as p t k and b d g, ignoring aspiration completely, and noting that b d g are only slightly voiced. Other sources on Faroese, as well as knowledge of related languages (e.g., Icelandic), suggest that the contrast for stops in Modern Faroese is one of aspiration rather than voicing. Marius Staksberg (2002, p.c.), Secretary of the Faroese Language Committee, explains that the opposition between the lenis and fortis stops is one of aspiration (pre- and post-aspiration, depending on context), not voicing (e.g., compare *bera* ['pe:ra] : *pera* ['p^{h}e:ra] and *bidda* ['pit:a] : *bitta* ['pi^{h}t:a]). He adds that the unaspirated series is traditionally transcribed as voiceless b d g (i.e., b̥ d̥ g̊). This may explain Lockwood's transcription method, and helps translate his chart of sounds for Faroese into the International Phonetic Alphabet system used here.

Description

Lockwood describes the voicing of simple stops as

> A feature of the pronunciation of many districts is the softening of [k p t] to [g b d] when occurring between vowels or medially before [l] (except [tl]) and [r]; occasionally [c] occurs between vowels and is likewise softened to [ɟ]. (p.22)

(3.29) Voicing of stops (p.22):

a.	taka	[tɛaga]	'to take'
b.	drepa	[dre:ba]	'to kill'
c.	sita	[si:da]	'to sit'

Discussion

Voicing for stops in Faroese is not contrastive (the opposition between the stops is manifested in aspiration) and therefore the voicing of one of the stop series[10] does not lead to neutralization with existing phonemes in the language.

Notes on Difference from Kirchner's list

Kirchner (1998) lists a second process for Faroese: in some dialects stops become breathy voiced between devoiced sounds and liquids. Kirchner's grammar source (Lockwood 1964) mentions nothing in his description of the language about voicing with breathiness, and therefore this process is omitted from my inventory.

3.34. Finnish

Language:	Finnish	*Dialect*:	Standard
Phylum:	Uralic, Finno-Ugric, Finno-Permic, Finno-Cheremisic, Finno-Mordvinic, Finno-Lappic, Baltic-Finnic, Finnic		
Inventory:	Lavoie	*Source*:	Sulkala and Karjalainen 1992
Lenition(s):	r → ɾ medially		
Type:	Flapping		
Classification:	Not Neutralization		
Difference:	One process is omitted		

[10] Lockwood's description of the stops series in Faroese does not make it clear whether he meant that the simple, or the aspirated, stops voice. If only one series voices the contrast between simple and aspirated is maintained. If, alternatively, both series voice but there is no change in aspiration (i.e., p → b and p^h → b^h) the contrast between phonemes is also maintained. The contrast would be obliterated if both simple and aspirated stops voiced to b d g, but judging by all the sources consulted on Faroese this was not the case.

Consonant Inventory

Table 46: Phonemic Consonant Inventory of Finnish

p		t				k	ʔ
(b)		d				(g)	
	(f)		s	(ʃ)			h
m		n				ŋ	
		r					
		l					
w					j		

Sounds in parentheses occur only in loanwords.

Description

In Finnish the trill is flapped word-medially.

Discussion

There are no flaps in the inventory of Finnish phonemes so the result of this lenition process cannot neutralize with existing sounds in the language.

Notes on Difference from Kirchner's list

Lavoie (1996), as cited in Kirchner (1998), lists an additional lenition process for Finnish: delete h, k word-finally. Her own grammar source, however, specifically calls this a relic of a morphological process no longer active in the language. There is no synchronic active deletion of word-final h or k in Finnish, and even when there was, it was not phonologically pervasive and therefore not an example of lenition.

3.35. French (Périgourdin)

Language:	French	*Dialect*:	Périgourdin
Phylum:	Indo-European, Italic, Romance, Italo-Western, Western, Gallo-Iberian, Gallo-Romance, Gallo-Rhaetian, Oïl, French		
Inventory:	Lavoie	*Source*:	Marshall 1984
Lenition(s):	1. t → d / V_V 2. d → ð / V_V		
Type:	1. Voicing		

	2. Spirantization
Classification:	1. Not Neutralization
	2. Not Neutralization
Trends:	Phonemic overlap between the two processes
Difference:	Sounds participating in both processes

Consonant Inventory

Table 47: Consonant Inventory of Périgourdin French

p			t				k
b			d				g
		f		s	ʃ		
		v		z	ʒ		
m			n			ɲ	
			ɾ				
			r				
			l			ʎ	
w	ɥ					j	

Description

In Périgourdin French there was a shift that voiced intervocalic t's (process 1) and spirantized intervocalic d's (process 2).

Discussion

The voicing of intervocalic t's resulted in intervocalic d's, and the spirantization of intervocalic d's resulted in ð. Thus, in the same environment where the t is voiced, the sound it would have neutralized with (d) is shifted, and no contrast between phonemes is obliterated. Since the ð is not phonemic, the second process did not lead to neutralization either.

Notes on Difference from Kirchner's list

Lavoie (1996), cited in Kirchner (1998), lists the intervocalic voicing of p and spirantization of b to β. Lavoie's grammar source (Marshall 1984), however, discusses these processes as affecting the alveolars (t and d) and says nothing of the bilabial p and b.

3.36. Gbɛya

Language:	Gbɛya	*Dialect*:	Not specified
Phylum:	Niger-Congo, Atlantic-Congo, Volta-Congo, North, Adamawa-Ubangi, Ubangi, Gbaya-Manza-Ngbaka, Central		
Inventory:	Lavoie	*Source*:	Smarin 1966
Lenition(s):	ʔ → Ø / V_V (optional)		
Type:	Loss		
Classification:	Not Neutralization		
Trends:	Phonetic, but not phonological, neutralization (ʔ/Ø)		

Consonant Inventory

Table 48: Phonemic Consonant Inventory of Gbɛya

p		t			k	kp	ʔ
b		d			g	gb	
^{m}b		^{n}d			^{ŋ}g	^{ŋ}gb	
^{ʔ}b		^{ʔ}d					
	f		s				h
	v		z				
m		n			ŋ	ŋm	
^{ʔ}m		^{ʔ}n					
		ɾ					
		l					
w	ʋ			j			

Description

Smarin describes the loss of ʔ as:

> The glottal stop has allophones [ʔ] and [zero] which alternate freely between vowels in close juncture and before vowels following pause or open juncture. In this latter, called initial, position [ʔ] occurs when something is being emphasized, but since there are so few words with initial [ʔV], this is not common. In medial position [ʔ] occurs in only the following words: yaʔa 'grandmother', naʔa 'mother', and bɔʔɔ used to intensify possession. (p.20)

Discussion

The intervocalic loss of ʔ constitutes phonetic neutralization with Ø. However, since this affects only 3 lexical items, no meaning distinctions could possibly be obliterated by this loss. Therefore, due to the extremely low functional load of the ʔ/Ø contrast in intervocalic context, the loss of ʔ is not neutralizing.

3.37. Georgian

Language: Georgian *Dialect*: Not specified
Phylum: South Caucasian, Georgian
Inventory: Kirchner *Source*: Aronson 1989
Lenition(s): v → w / C_ and in coda
Type: Gliding
Classification: Not Neutralization
Difference: One process omitted; context corrected for included process

Consonant Inventory

Table 49: Consonant Inventory of Georgian

p^h		t^h	ts^h		c^h	k^h		
$p^ʔ$		$t^ʔ$	$ts^ʔ$		$c^ʔ$	$k^ʔ$	$q^ʔ$	$qχ^ʔ$
b		d	dz		ɟ	g		
			s	ʃ			χ	
							$χ^ʔ$	
(β)	v		z	ʒ			ʁ	
m		n						
		ɾ						
		l						
		(ɬ)						
(w)								

Aronson indicates that β v w are variations of the same phoneme, and l ɬ are as well.

Description

Aronson describes the gliding of v in Georgian as:

> Georgian v shows fluctuation in its pronunciation from [w] to [v] ... [v] occurs most commonly initially and intervocalically, [w] after consonants and at the end of syllables. It can also be realized as a bilabial voiced fricative [β] and, before voiceless consonants, as [f]. (p.27)

Discussion

The gliding of v in Georgian is not neutralizing because all cases of w in the language are results of this lenition process, so neutralization with existing phonemes is not possible.

Notes on Difference from Kirchner's list

Kirchner (1998) lists two lenition processes for Georgian ($q^{ʔ}$ ~ $qχ^{ʔ}$ ~ χ and v - w). The first process is not found in Kirchner's grammar source (Aronson 1989) except as a note that all three sounds are written with the same orthographic symbol. While this may be an indication of a past, or even present, alternation, there is no reason to believe that is the case from Aronson's description, especially since he lists the three sounds separately in his chart (in contrast to the way he lists β v w or l ɬ to show alternations). For this reason I omit this process from my inventory. I also correct the context of v - w to more closely resemble Aronson's description.

3.38. Gitksan

Language:	Gitksan	*Dialect*:	Not specified
Phylum:	Penutian, Tsimshian		
Inventory:	Lavoie	*Source*:	Hoard 1978
Lenition(s):	p t ts k k^w q → b d dz g g^w ɢ before sonorants		
Type:	Voicing		
Classification:	Not Neutralization		
Difference:	Process description corrected		

Consonant Inventory

Table 50: Phonemic Consonant Inventory of Gitksan

p	t		ts		k	kʷ	q	ʔ
pˀ	tˀ	tlˀ	tsˀ		kˀ	kʷˀ	qˀ	
			s		x	xʷ	χ	h
m	n							
mˀ	nˀ							
	l							
	lˀ							
	ɬ							
w				j				
wˀ				jˀ				

Description

On the distribution of sounds in Gitksan Hoard writes that "before any voiced sonorant, plain stops and affricates are voiced; elsewhere they are voiceless" (p.114).

Discussion

There is little discussion of the functional considerations of the voicing process in Gitksan, but the phoneme chart makes it clear that voicing is not contrastive in the language. The voiced stops surface only as results of this voicing process and therefore do not neutralize with existing sounds.

Notes on Difference from Kirchner's list

Lavoie (1996), listed in Kirchner (1998), describes the lenition process in Gitksan as p - b / _V. I follow her language source (Hoard 1978) in describing the voicing as affecting all stops and affricates in the language before sonorants.

3.39. Gojri

Language: Gojri *Dialect*: Not specified
Phylum: Indo-European, Indo-Iranian, Indo-Aryan, Central zone, Rajasthani, Unclassified
Inventory: Kirchner *Source*: Sharma 1979
Lenition(s): CC - C: (extra long geminates reduce to plain geminates)

Type: Shortening (but not actual degemination)
Classification: Not Neutralization

Consonant Inventory

Table 51: Phonemic Consonant Inventory of Gojri

p		t		ʈ	c	k	q	
pʰ		tʰ		ʈʰ	cʰ	kʰ		
b		d		ɖ	ɟ	g		
	f		s		ç	x		h
			z			ɣ		
m		n		ɳ				
				ɽ				
		r						
		l		ɭ				
w					j			

Description

The long stops in Gojri alternate with their extra long counterparts intervocalically after certain vowels.

Discussion

Each stop (voiceless, aspirated, and voiced) can occur in 1 of 3 lengths: singleton (C), long (C:), and extra long (CC). The singletons can occur in all positions, and contrasts with the longer versions (3.30).

(3.30) Contrast between short and long consonants (p.32):

a.	/bədɛ/	'increases'	/bəddɛ/	'ties'
b.	/əɖa/	'make something fly'	/aɖɖa/	'bus stop'

The long and extra long varieties are in complementary distribution: the long stops (C:) are found intervocalically after i e ɛ a o ɔ, and the extra long stops (CC) are found after ɨ ə u intervocalically. Hence, in Gojri there is a contrast between the short and the long stops (C/C:), but there is no contrast between the long and the extra-long stops (C:/CC). Therefore, the alternation between the long and the extra-long stops (C:-CC) does not neutralize an existing contrast in the language.

3.40. Gooniyandi

Language:	Gooniyandi	*Dialect*:	Not specified
Phylum:	Australian, Bunaban		
Inventory:	Lavoie	*Source*:	McGregor 1990
Lenition(s):	Stops voice word internally following nasals and laterals (& elsewhere)		
Type:	Voicing		
Classification:	Not Neutralization		
Difference:	Process description corrected		

Consonant Inventory

Table 52: Phonemic Consonant Inventory of Gooniyandi

p	t̪	t	ʈ	c	k
m	n̪	n	ɳ	ɲ	ŋ
		ɾ	ɭ	ʎ	
		l			
w			ɻ	j	

McGregor adds to this inventory that "the glottal stop [ʔ] occurs in a few interjections, such as [nʔn] 'yes' and is occasionally used in elicited speech to mark morpheme boundaries" (p.37).

Description

McGregor reports that Gooniyandi has a single stop series, with voiced and voiceless allophones occurring in various environments. A context where stops are consistently voiced is word internally, following nasals and laterals (3.31a, b, c). Also, stops tend to voice in word-initial stop-liquid clusters (3.31d).

(3.31) Internal voicing (p.49-50):

a.	kampa	['kambʌ]	'water'
b.	jilpa	['ji.lbʌ]	'for ever, completely'
c.	wiɲciɽki	[ˌwiɲ'ɟɜɽki]	'top of back'
d.	pɽijanti	['bɻeiændi]	'in retaliation'

McGregor adds that unreleased stops may occur syllable finally, and voicing from the preceding vowel may extend into the period of occlusion. Also in word medial stop-stop clusters there is assimilation to the preceding member's voicing.

Discussion

Voicing is not contrastive in Gooniyandi so the voiceless stops, and their voiced counterparts, are allophones of the same phonemes. Therefore, the voicing of stops in various contexts does not neutralize any contrasts in the language.

Notes on Difference from Kirchner's list

Lavoie (1996), as reprinted in Kirchner (1998), describes the voicing in Gooniyandi as p → b / V_V. Her language source (McGregor 1990) however, reports that all the stops participate in the voicing process. Furthermore, although voicing occurs in many other contexts, intervocalically it is most rare. The author specifically says that both voiced and voiceless allophones may occur intervocalically, but the second is more common in that position.

3.41. Gothic

Language: Gothic *Dialect*: Not specified
Phylum: Indo-European, Germanic, East
Inventory: Lavoie *Source*: Bennett 1980
Lenition(s): b d g → β ð ɣ medially
Type: Spirantization
Classification: Not Neutralization

Consonant Inventory

Table 53: Phonemic Consonant Inventory of Gothic

p	t			k
				k^w
b	d			g
ɸ	θ	s		x
				x^w
		z		
m	n			
	r			
	l			
			j	

Description

In Gothic the voiced stops spirantize word medially after a vowel or a diphthong.

Discussion

Bennett is very short on details but he does provide an inventory of phonemes in which the voiced fricatives are not included. According to this, spirantization for voiced stops is not contrastive, and the only β ð ɣ that surface in Gothic are results of the process in question. Therefore no contrast is neutralized by the spirantization of b d g.

3.42. Greenlandic (West)

Language:	Greenlandic	*Dialect*:	West
Phylum:	Eskimo-Aleut, Eskimo, Inuit		
Inventory:	Lavoie	*Source*:	Fortescue 1984
Lenition(s):	q → χ or ʁ / V_V		
Type:	Spirantization		
Classification:	Not Neutralization		

Consonant Inventory

Table 54: Phonemic Consonant Inventory of West Greenlandic

p		t					k	q	(ʔ)
						ɟ	g		
	(f)		s	ʃ					(h)
	v								
m		n					ŋ	(ɴ)	
								ʀ	
		l							
w					ɻ				

Sounds in parentheses are "marginal": ɴ is rare, and f ʔ h occur only in loan words.

Description

In intervocalic position the voiceless uvular stop q has the following two variants: χ or ʁ.

Discussion

The voiced and voiceless spirants that surface as the intervocalic variants of the uvular stop are not listed in the phonemic inventory. Therefore, the spirantization of q does not neutralize distinction with other phonemes.

Additional Comments

The inventory of phonemes for West Greenlandic seems rather asymmetrical, especially where the voiceless and voiced stops are concerned. However, classifying q - χ ʁ as not neutralizing is not affected by the possibility that the list of stop phonemes may be incomplete.

3.43. Guayabero

Language: Guayabero *Dialect*: Not specified
Phylum: Guahiban
Inventory: Lavoie *Source*: Keels 1985
Lenition(s): d → θ syllable-finally
Type: Spirantization and devoicing
Classification: Not Neutralization

Consonant Inventory

Table 55: Phonemic Consonant Inventory of Guayabero

p	t		tʃ		k	?
b	d					
ɸ	(θ)	s			x	h
(β)						
m	n					
	r					
	l					
w				j		

The sounds in parentheses (θ and β) are in complementary distribution with d and w.

Description

Keels reports that in Guayabero θ is the syllable-final variant of d.

(3.32) Example of d - θ (p.84):
miadham [miaθhʌm] 'you dig'

Discussion

Keels reports that d and θ are in complementary distribution and that θ is an allophone of d and is not a phoneme in the language. Since θ only surfaces in the language as a result of this spirantization, no contrast is neutralized by this process.

3.44. Gujarati

Language:	Gujarati *Dialect*: Not specified
Phylum:	Indo-European, Indo-Iranian, Indo-Aryan, Central zone, Bhil
Inventory:	Kirchner *Source*: Cardona 1965
Lenition(s):	1. b^{h} d^{h} $ɖ^{h}$ g^{h} ~ Hβ Hð Hɽ Hɣ / V_V (free variation)
	2. ɖ ~ ɽ / _# (free variation)
Type:	1. Spirantization
	2. Flapping
Classification:	1. Not Neutralization
	2. Not Neutralization
Trends:	Flapping patterns with spirantization; also apparent merger (ɖ/$ɖ^{h}$)
Difference:	Sounds included in the alternations are corrected

Consonant Inventory

Table 56: Phonemic Consonant Inventory of Gujarati

p	t		tʃ	ʈ		k	
b	d		dʒ	ɖ		g	
		s	ʃ				
		z					ɦ
m	n			ɳ			
	ɾ						
	l			ɭ			
w					j		

Description

Process 1 involves the spirantization of aspirated voiced stops. Note that Cardona does not include the aspirated voiced stops in his inventory of phonemes for the language, suggesting that aspiration on voiced stops is not contrastive. A discussion of this point can be found below. Cardona reports that the spirant variants β ð ɽ ɣ occur intervocalically following a murmured vowel (designated by H), and vary freely with the aspirated voiced variants.

(3.33) Examples of b^h d^h ɖ^h g^h ~ Hβ Hð Hɽ Hɣ (p.23):

a.	[wag^{h}əɳ]	~	[waHɣəɳ]	'tigress'
b.	[sud^{h}i]	~	[suHði]	'until'
c.	[ab^{h}ar]	~	[aHβar]	'thanks'

In Gujarati ɖ and ɽ vary freely in final position, which is Process 2.

(3.34) Example of ɖ ~ ɽ (p.24):

[ɟ^{h}aɖ] ~ [ɟ^{h}aɽ] 'tree'

Discussion

Spirantization of voiced stops is not contrastive in Gujarati so β ð ɽ ɣ only occur in the language as part of the lenition processes described here. Therefore there is no neutralization with existing sounds in the language, and no contrast is obliterated by process 1.

On the face of it, it would seem that the result of process 2 would neutralize with the result of process 1, obliterating the contrast between ɖ^h and ɖ. However, this is not the case for three reasons: first, in Cardona's description of the language there is no contrast between voiced and voiced aspirated stops, so neutralization of this contrast is not possible. Second, the two processes occur in different contexts. Third, the aspirated stops alternate with spirants preceded by breathiness, while the plain ɖ alternates with a plain ɽ. That is, if aspiration of voiced stops was in fact contrastive, this contrast would be maintained in the breathiness of the preceding vowel, shifting the ɖʰ/ɖ opposition to Hɽ/ɽ. Hence, the second process is also not neutralizing.

Additional Comments

Cardona does not include aspiration, either for the voiced or voiceless sounds, in his inventory of phonemes. Despite Cardona's description of the grammar there is reason to believe that aspiration in Gujarati is contrastive, although it isn't clear without further investigation whether aspiration is contrastive for only the voiceless or the entire stop series. For the purposes of my investigation, however, this question is significant but not crucial. That is, whether aspiration is contrastive or not, it is preserved in the alternations above.

Notes on Difference from Kirchner's list

In his list of processes for Gujarati, Kirchner (1998) has the retroflex stop, both aspirated and not, alternating with the alveolar flap. If this were true, both processes would have caused neutralization not only with each other but also with existing phonemes in the language. Kirchner's grammar source (Cardona 1965), however, specifies that the retroflex stop alternates with the retroflex flap, which is not phonemic. Furthermore, he specifies that the aspirated stops alternate with a pre-breathy spirant as I list in my version of process 1.

3.45. Haitian Creole

Language:	Haitian Creole	*Dialect*: Not specified
Phylum:	Creole, French based (Indo-European, Italic, Romance)	
Inventory:	Lavoie	*Source*: Tinelli 1981
Lenition(s):	1. ij → i / _#	
	2. ʒ → j / _#	
Type:	1. Loss	
	2. Gliding	
Classification:	1. Not Neutralization	
	2. Not Neutralization	
Trends:	Phonetic, but not phonological, neutralization (process 1)	

Consonant Inventory

Not provided in the source

Description

Process 1 is listed by Lavoie (1966) as the deletion of final glides, but according to her source (Tinelli 1981) the only glide that can occur in this context is j.

(3.35) Formalization of glide deletion rule (p.126):
Fr. j > null Hn. in word final position after i

Process 2, gliding of the voiced sibilant, refers to Tinelli's statement that in some cases Haitian j derives from French z or ʒ.

Discussion

Process 1 is based on Tinelli's description of the change, or rather difference, between French and Haitian Creole, which he calls the deletion of word final j after i. He adds that there are very few actual words that are affected by this rule, providing the only examples.

(3.36) Words affected by glide deletion (p.126):

French	Haitian	
fij	fi	'girl'
famij	fami	'family'

Regardless of whether the glide-deletion is an active phonological process in the language or not (see next section for discussion of this point) there is no possibility of neutralization resulting from the deletion of j because it affects only two words in the language.

Process 2 also describes a difference between French and Haitian which, again, affects very few words: "This change is rare, since few French words ending in a ʒ have been adopted by its creoles" (p.126). Just as with process 1, neutralization is not possible due to the fact that very few words are affected by this phenomenon.

Additional Comments

Both processes listed by Lavoie for Haitian Creole are descriptions of differences between French and Haitian, and there is no reason to believe that either of these phenomena was ever an active process in Haitian phonology. Most probably, for example, fij was never the word for girl in Haitian that was actively changed to fi by dropping the word final j. Haitian Creole may be French-based, but that does not imply that speakers of Haitian internalize the French lexicon which then undergoes "creolization" by phonological rules, as would be implied by calling the above processes lenitions in Haitian.

3.46. Hebrew (Tiberian)

Language:	Hebrew *Dialect*: Tiberian
Phylum:	Afro-Asiatic, Semitic, Central, South, Canaanite
Inventory:	Lavoie & Kirchner *Source*: Malone 1993
Add. Sources:	Prince 1975, Lowenstamm and Kaye 1986, Coetzee 1999
Lenition(s):	1. p b t d k g → f v θ ð x ɣ / V_ 2. Word-final degemination
Type:	1. Spirantization 2. Degemination
Classification:	1. Not Neutralization 2. Not Neutralization
Difference:	A duplicate process is omitted; process 1 from Lavoie, 2 from Kirchner

Consonant Inventory

Table 57: Consonant Inventory of Tiberian Hebrew (Prince 1975)

p		t				k	q		ʔ
b		d				g			
	(f)	(θ)	s	ʃ	ç	(x)		ħ	h
	(v)	(ð)	z			(ɣ)		ʕ	
m		n							
		r							
		l							
w					j				

The alveolars t and s have emphatic variants; q is emphatic. Sounds in parentheses are not phonemic.

Description

Process 1 is the postvocalic spirantization of the non-emphatic stops p b t d k g.

Process 2 is described by Malone (1993: 73) as the "simplification of would-be geminate clusters in word-final position", as illustrated in (3.37).

(3.37) Word-final degemination (Malone 1993: 73):
[qal] 'light (masc.)' vs. [qallɔɔ] 'light (fem.)'

Discussion

Process 1 is obviously allophonic because, as Prince (1975: 9) points out, the spirantization of non-emphatics "is the only source for these spirants". Therefore the spirants that alternate with the non-emphatic stops do not affect other phonemes in the language and do not lead to any neutralization of contrasts.

Malone (1993) does not provide much information about the contrasts in Tiberian Hebrew. Prince (1975: 11) reports that "geminates arise through derivation, suffixal inflection, and phonological assimilation". If all gemination is indeed derivational, it cannot be contrastive. Therefore degemination in any environment cannot be neutralizing: it does not obliterate contrast between phonemes because no such contrast depends solely on length.

Notes on Difference from Kirchner's list

Lavoie (1996), cited in Kirchner (1998), lists the spirantization process in Tiberian Hebrew, which is the 1st process above. In his inventory Kirchner adds two more processes for this language using the same language source as Lavoie (Malone 1993): the first is the exact same spirantization of non-emphatics already listed in his inventory, and the second is the degemination process (process 2 above). I omit the duplicate spirantization process.

3.47. Icelandic

Language: Icelandic *Dialect*: Not specified
Phylum: Indo-European, Germanic, North, West Scandinavian
Inventory: Kirchner *Source*: Thrainsson 1978
Lenition(s): pp tt kk - ʰp ʰt ʰk
Type: Preaspiration
Classification: Not Neutralization

Consonant Inventory

Not provided in the source

Description

Thrainsson discusses an alternation between pp tt kk and ^{h}p ^{h}t ^{h}k. He presents an example of a similar diachronic rule that is illustrated in words that, although they do not have an active alternation, do display a difference between the spelling and pronunciation (3.38). Examples of an active alternation are provided in (3.39).

(3.38) Evidence of a diachronic pp tt kk - ^{h}p ^{h}t ^{h}k (p.6):

a.	kappi	[k^{h}a^{h}pɪ]	'hero'
b.	θakka	[θa^{h}ka]	'thank'
c.	hattur	[ha^{h}tʏr]	'hat'

(3.39) Evidence of an active alternation (p.8):

	Fem. sg.			Neut. sg.	
a.	feit	[fei:t^h]	'fat'	feitt	[fei^{h}t]
b.	ljot	[ljou:t^h]	'ugly'	ljott	[ljou^{h}t]
c.	sæt	[sai:t^h]	'sweet'	sætt	[sai^{h}t]

The synchronic preaspiration process also affects geminates created by assimilation as well as neologisms, loanwords, and Icelandic accent.

Discussion

According to Thrainsson, preaspiration in itself is not phonemic in the language and is only derived by rules. However, there is a second rule that preaspirates p t k before l or n (3.40).

(3.40) Preaspiration occurring before l, n (p.16):

a.	epli	[ɛ^{h}pli]	'apple'	b.	opna	[ɔ^{h}pna]	'open'
c.	ætla	[ai^{h}tla]	'intend'	d.	vetni	[vɛ^{h}tnɪ]	'hydrogen'
e.	ekla	[ɛ^{h}kla]	'lack'	f.	vakna	[va^{h}kna]	'wake up'

It appears that the preaspirated ^{h}p ^{h}t ^{h}k can be derived from one of two processes: the degemination and preaspiration of pp tt kk, and the preaspiration of p t k before l or n. The question now is whether this can occur in similar environments thereby neutralizing the contrast between pp tt kk and p t k.[11]

[11] It is not clear from the source that there actually is a contrast between geminates and singletons. If such contrast does not exist in the language there is no possibility of

The preaspiration of singletons cannot occur either word finally or intervocalically. In fact, the only way neutralization would be possible is if clusters such as geminate + l or n existed in the language creating a situation where, e.g., ʰpl could result from either ppl or pl. But, based on the fact that Thrainsson claims all cases of preaspirated stops followed by l or n are cluster-initial singletons, this problem is apparently avoided.

I conclude that the preaspiration and degemination of pp tt kk does not result in neutralization because the only other preaspirated stops in the language occur in an environment where geminates do not. In other words, it is clear from the context whether a preaspirated stop is a representation of a geminate or of a singleton followed by l or n, and the contrast which possibly (or rather, probably) exists between geminates and singletons in the language is maintained.

3.48. Italian (Southern)

Language:	Italian	*Dialect*: Southern	
Phylum:	Indo-European, Italic, Romance, Italo-Western, Italo-Dalmatian		
Inventory:	Kirchner	*Source*: Oftedal 1985	
Lenition(s):	p t k → b̥ d̥ g̊ non-initially		
Type:	Voicing		
Classification:	Not Neutralization		

Consonant Inventory

Not provided in the source

Description

Oftedal writes that "in many dialects of Central and Southern Italy, [p t k] are weakened to half-voiced [b̥ d̥ g̊] between vowels, even in word juncture" (p.35).

(3.41) Examples of partial voicing (p.36):

a.	parte	[ˈparte]	'part':	di parte	[di ˈb̥arte]	'of a part'
b.	terra	[ˈterra]	'land':	la terra	[la ˈd̥erra]	'the land'
c.	carne	[ˈkarne]	'meat':	di carne	[di ˈg̊arne]	'of meat'

---

neutralization resulting from the lenition in question. If this contrast does exist, however, I am concerned with whether it would be preserved after preaspiration.

Discussion

Although an inventory of phonemes is not provided in the source, the author does address the issue of contrasts, and asserts that the partially voiced stops b̥ d̥ g̊ remain distinct from b d g: "Sometimes the weakened phones are voiced, but Weinrich (1958: 51) and Tekavčić (1972: 175) agree that these voiced stops remain distinct from the voiced phonemes /b d g/" (p.36). Hence neutralization between the lenited p t k and the voiced b d g in the language is avoided.

3.49. Kabardian (Terek)

Language: Kabardian *Dialect*: Terek
Phylum: North Caucasian, Northwest, Circassian
Inventory: Lavoie *Source*: Colarusso 1988
Lenition(s): ts' - s'
Type: Spirantization
Classification: Not Neutralization

Consonant Inventory

Table 58: Consonant Inventory of Terek Kabardian

p		t	ts			c		q		ʔ
p'		t'	ts'			c'		q'		
							kʷ	qʷ		ʔʷ
							kʷ'	qʷ'		
b		d	dz			ɟ	gʷ			
	f		s	ʃ	ɕ	ç		χ	ħ	h
	f'			ʃ'						
							xʷ	χʷ		
	v		z	ʒ	ʑ	j		ʁ	(ʕ)	
								ʁʷ		
m		n								
		r								
		ɬ		ɮ						
		ɬ'								
w										

The language source is a detailed analysis of the diachronic changes from East Circassian to its various daughter languages so the inventory is probably not a list of distinct sounds in any one language or dialect.

Description

In the Terek dialect of Kabardian the ejective alveolar affricate ts' was spirantized to s'. The author also notes that the result is a sound unique to the Terek dialect.

Discussion

Colarusso's objective is to describe the diachronic changes in the East Circassian languages, and how the various contrasts in the original language manifest themselves in modern dialects. The result of the diachronic spirantization process in the Terek dialect of Kabardian (s') is not included in the inventory of the proto-language. The fact that the s' is a sound that only exists in Terek provides further evidence that it was not phonemic in the proto language. This suggests that contrast was maintained in the shift from the affricate to the spirant, and no neutralization resulted from this process.

3.50. Kagate

Language:	Kagate	*Dialect*:	Spoken in village of Phedi, Saipu-Sabra Panchayat Ramechhap
Phylum:	Sino-Tibetan, Tibeto-Burman, Himalayish, Tibeto-Kanauri, Tibetic, Tibetan, Central		
Inventory:	Kirchner	*Source*:	Hoehlig & Hari 1976
Lenition(s):	1. b ~ β / V_V 2. g → lenis / V_V 3. r → ɾ / V_V		
Type:	1. Spirantization 2. Unclear 3. Flapping		
Classification:	1. Not Neutralization 2. Not Neutralization 3. Not Neutralization		
Difference:	Source reference is corrected		

Consonant Inventory

Table 59: Phonemic Consonant Inventory of Kagate

p	t	ts	ʈ		k	
p^h	t^h	ts^h	$ʈ^h$		k^h	
b	d	dz	ɖ		g	
		s				h
m	n				ŋ	
	r					
	r^h					
	l					
	l^h					
w				j		

Description

Process 1 is the intervocalic fluctuation between b and β. In the grammar the authors use a vowel + h combination (e.g., ah) and a diacritic (e.g., a̤) to indicate a lax vowel.

(3.42) b ~ β / V_V (p.15):

a.	/libi/	[iːβi / iːbi]	'grandmother'
b.	/prjahbe/	[prjæβe / prjæbe]	'echo'
c.	/mehbe/	[me-βe / me-be]	'do not'

Process 2 is the alternation between g and its lenis counterpart intervocalically. Examples of words where this occurs are provided, but how the lenition is manifested is never explained.

(3.43) Intervocalic lenition of g (p.18):

a.	/p^hahge/	[p^ha̤/ge]	'measles'
b.	/kohga/	[ko̤/ga]	'fire place'

Process 3 is the intervocalic flapping of the trill.

(3.44) r → ɾ / V_V (p.19):

a.	/tari/	[ta\ri]	'axe'
b.	/tihriŋ/	[ti̤/riŋ]	'today'
c.	/guhri/	[gṳri]	'cat'

Discussion

All three processes result in sounds that do not already exist in the inventory of the language (β, lenis ɡ, ɾ) and therefore do not neutralize any contrasts.

Notes on Difference from Kirchner's list

Kirchner (1998) lists the source for his data on Kagate as 'Hoehlig 1978'. I stay consistent in using the cited grammar, but correct the reference to 'Hoehlig & Hari 1976'.

3.51. Kaliai-Kove (Kandoka-Lusi)

Language: Kaliai-Kove *Dialect*: Kandoka-Lusi
Phylum: Austronesian, Malayo-Polynesian, Central-Eastern, Eastern Malayo-Polynesian, Oceanic, Western Oceanic, North New Guinea, Ngero-Vitiaz, Ngero, Bariai
Inventory: Kirchner *Source*: Counts 1969
Lenition(s): ɡ ~ ɣ in free variation
Type: Spirantization
Classification: Not Neutralization
Difference: Sounds participating in the process are corrected

Consonant Inventory

Table 60: Phonemic Consonant Inventory of Kandoka-Lusi Kaliai-Kove

p	t		k	
^{m}b	^{n}d		^{ŋ}g	
		s		h
β			ɣ	
m	n		ŋ	
	ɾ			
	r			
	l			
w				

Description

Counts describes the velar fricative phoneme ɣ as having two allophones, ɣ and ɡ, which are in free variation.

Discussion

The velar voiced stop g, which is in free variation with ɣ, is not included in the inventory of phonemes for Kaliai-Kove. This means that g only surfaces in the language as part of the g ~ ɣ alternation. In addition to this, the author is concerned with the ᵑg/g opposition and writes that the two sounds are "allophones of separate phonemes despite their phonetic similarity and the restricted occurrence of ᵑg" (p.18). The existence of minimum pairs such as (3.45) illustrates this point.

(3.45) Minimum pair that shows ᵑg and g are contrastive (p.18):

a.	/paɣao/	[phaga˧o]	'to me'
b.	/paᵑgao/	[phaᵑga˧o]	a man's name

Thus, the free variation of ɣ ~ g does not lead to the neutralization of any contrasts.

Notes on Difference from Kirchner's list

Kirchner (1998) adds b ~ β to the above spirantization process in Kaliai-Kove. I follow Kirchner's grammar source (Counts 1969) in listing only ɣ ~ g, as Counts specifically notes that the phoneme /β/ exhibits a single allophone [β].

3.52. Kanakuru

Language:	Kanakuru	*Dialect*:	Not specified
Phylum:	Afro-Asiatic, Chadic, West, A, A.2, Tangale, Dera		
Inventory:	Kirchner	*Source*:	Newman 1974
Lenition(s):	p t k → w ɹ ɰ / V_V, _# with lexical exceptions		
Type:	Gliding		
Classification:	Not Neutralization		
Difference:	Process description is corrected		

Consonant Inventory

Table 61: Consonant Inventory of Kanakuru

p	t			k
				k^{w}
b	d		ɟ	g
	d^{w}			g^{w}
^{m}b	^{n}d		^{ɲ}ɟ	^{ŋ}g
				^{ŋ}g^{w}
ɓ	ɗ			
ɓ^{w}	ɗ^{w}			
		ʃ		
		ʃ^{w}		
m	n		ɲ	ŋ
	ɬ	(ɮ)		
w	ɹ		j	ɰ

Description

Newman describes the gliding in Kanakuru as:

> The most striking feature of Kanakuru phonology and morphophonemics is the pervasive alternation of voiceless stops and the sonorants w, ɹ, and ɰ. Historically this alternation was brought about by the massive weakening of stops in medial position and subsequent adjustments and analogical leveling in final position. (p.4)

This process is blocked in a few lexical items.

Discussion

Newman does not isolate the phonemes in the language's inventory, but he does, in his description, specify that p/w, t/ɹ, and k/ɰ are allophones of the same phonemes. Hence, the gliding alternation is allophonic and does not lead to neutralization of any contrasts.

Notes on Difference from Kirchner's list

Kirchner (1998) lists the lenition for Kanakuru as p t k - w/p' ɾ ɣ. I follow Kirchner's grammar source (Newman 1974) in listing the alternation as one of gliding rather than spirantization (p t k - w ɹ ɰ).

3.53. Kannada

Language:	Kannada	*Dialect*:	Not specified
Phylum:	Dravidian, Southern, Tamil-Kannada		
Inventory:	Lavoie	*Source*:	Chisum 1975
Add. Sources:	Andronov 1969 (language source for Chisum), Diffloth 1975		
Lenition(s):	p t k → b d g / V_V		
Type:	Voicing		
Classification:	Not Neutralization		
Difference:	1st process is corrected, 2nd and 3rd processes are omitted		

Consonant Inventory

Table 62: Consonant Inventory of Kannada (Andronov 1969)

p	t	ts		ʈ		k	
pʰ	tʰ	tsʰ		ʈʰ		kʰ	
b	d	dz		ɖ		g	
bʰ	dʰ	dzʰ		ɖʰ		gʰ	
		s	ʃ	ʂ			h
m	n			ɳ	ɲ	ŋ	
	r						
	l			ɭ			
w					j		

Description

p t k are voiced when they find themselves in intervocalic position upon prefixation (3.46). Intervocalic voicing applies generally (3.47a, b), but with lexical exceptions (3.47c, d).

(3.46) p t k → b d g upon prefixation (Andronov 1969: 24):

a.	hale	+	kannaɖa	=	haɭegannaɖa	'Old Kannada'
b.	keɭa	+	tuʈi	=	keɭaduʈi	'lower lip'
c.	hu:	+	pu:ʈʈi	=	hubuʈʈi	'flower basket'

(3.47) Additional intervocalic voicing (Andronov 1969: 51):

a.	esetu	(esedu)	'to throw'
b.	kusitu	(kusidu)	'to sink'
c.	kalitu		'to learn'
d.	kaɭitu		'to ripen'

Discussion

The intervocalic voicing of p t k would be neutralizing if voicing was contrastive in Kannada. The inventory of sounds provided by Andronov does not indicate which are phonemic, and other sources on the language are scarce. However, one can look at other Dravidian languages, especially Southern ones, for help with this question. As it turns out, in general Dravidian languages do not have a voicing contrast, but they do have a singleton/geminate contrast (something that is also missing from the inventory of sounds for this language). If Kannada patterns with other languages in its family, and there is no reason why it shouldn't, voicing in this language is also not contrastive and therefore the intervocalic voicing of p t k is not neutralizing.

Additional Comments

There is a possibility that in some Dravidian languages the geminate/singleton contrast shifted to one of voicing, in which case a change in voicing would neutralize the original geminate/singleton contrast. However possible degemination in Dravidian languages lagged behind intervocalic voicing and is therefore unlikely to interfere with the classification of this process as not neutralizing (Hans Henrich Hock, p.c.).

Notes on Difference from Kirchner's list

Lavoie (1996), as cited in Kirchner (1998), lists three lenition processes for Kannada: k → g / V_V, delete glide / V_V, and word-initial p > h. However, neither her source (Chisum 1975) nor Chisum's primary grammar source (Andronov 1969) mentions the 2nd or 3rd processes on Lavoie's list. For this reason both are omitted from my inventory. The voicing process is corrected to include p and t, as reported in the primary grammar on the language.

3.54. Kanuri

Language:	Kanuri	*Dialect*:	Not specified
Phylum:	Nilo-Saharan, Saharan, Western, Kanuri		
Inventory:	Lavoie	*Source*:	Lukas 1967
Lenition(s):	b g → β ɣ between non-nasal voiced sounds in some dialects		
Type:	Spirantization		
Classification:	Not Neutralization		
Difference:	Affected sounds and the context are corrected		

Consonant Inventory

Table 63: Consonant Inventory of Kanuri

p		t		tʃ			k	ʔ
b		d		dʒ			g	
ɸ	f		s	ʃ			x	h
β			z					
m		n				ɲ	ŋ	
		r						
		l			ɭ			
w						j		

Description

"β is always derived from b; β is used between two voiced sounds, except after nasals, where b is retained" (p.3). In some cases b is retained such as in some reduplications (3.48a) and borrowings (3.48b). The g is slightly spirantized between two voiced sounds (also with some exceptions).

(3.48) Exceptions to b - β alternation (p.3):

a. baba˥˩ŋŋi˨n 'I beat often'

b. labu˧dda˩ 'doubtlessly' (from Arabic la bu˧dda˩ 'there is no doubt')

(3.49) Example of intervocalic ɣ (p.2):

sə˧diga˩ sə˧diɣa˩ 'bottom'

Discussion

The spirant variant of b (β) is specified by the author as existing in the language only as a derivation of b. The spirant variant of g (ɣ) is not included in the inventory of sounds for the language at all. Clearly neither β nor ɣ are phonemes in the language and therefore no contrasts are neutralized by the spirantization process in question.

Additional Comments

It is a little strange that β is included in the inventory despite the fact that it is recognized as an allophone of b and not a phoneme, but ɣ is nowhere to be found. This is probably just a matter of convention and possibly is related to the fact that g, according to the author, is only "slightly fricativized" in a voiced environment and possibly is not fricativized enough to be considered ɣ.

Notes on Difference from Kirchner's list

Lavoie (1996), as presented in Kirchner (1998), lists the lenition process for Kanuri as "b g → v ɣ / V_V". I follow her language source (Lukas 1967) in transcribing the spirant variant of b as β, especially as Lukas writes "Kanuri has no labio-dental v" (p.3). I also correct the environment where this rule applies to more accurately represent Lukas' description.

3.55. Karao

Language:	Karao	*Dialect*:	Not specified
Phylum:	Austronesian, Malayo-Polynesian, Western Malayo-Polynesian, Northern Philippine, Northern Luzon, South-Central Cordilleran, Southern Cordilleran, Pangasinic, Benguet, Ibaloi-Karao		
Inventory:	Kirchner	*Source*:	Brainard 1994
Lenition(s):	p t d tʃ dʒ g^{w} q → ɸ θ l ɾ j w χ / V_V		
Type:	Spirantization		
Classification:	Not Neutralization		
Trends:	Flapping patterns with spirantization		

Consonant Inventory

Table 64: Consonant Inventory of Karao

p	t		tʃ		k	q	ʔ
b	d		dʒ		g		
					g^{w}		
ɸ	θ	s				χ	
m	n				ŋ		
	ɾ						
	l						
w				j			

Description

Brainard writes that upon affixation there is an alternation in intervocalic position between the less sonorous stops and affricates (p t d tʃ dʒ g^{w} q) and their more sonorant counterparts (ɸ θ l ɾ j w χ). He adds that "those consonants that alternate with each other under such conditions are identical to those that are in complementary distribution with each other" (p.6).

Discussion

The complementary distribution between the stops/affricates and their more sonorous counterparts is not perfect as the more sonorous phonemes that do not result from the alternation in question may sometimes, although rarely, occur in intervocalic position. But in general the results of the spirantization process listed here appear in different contexts than existing similar sounds and therefore neutralization is avoided. What little neutralization there is from the rare exceptions to the complementary distribution between the stops and fricatives in the language is extremely limited.

3.56. Kashmiri

Language: Kashmiri *Dialect*: Not specified
Phylum: Indo-European, Indo-Iranian, Indo-Aryan, Northwestern zone, Dardic, Kashmiri
Inventory: Kirchner *Source*: Kachru 1969
Lenition(s): ɖ → ɽ / V_V, _#
Type: Flapping
Classification: Not Neutralization
Difference: Process description is corrected

Consonant Inventory

Table 65: Phonemic Consonant Inventory of Kashmiri

p	t	ts	tʃ	ʈ		k	
p^{h}	t^{h}	ts^{h}	tʃ^{h}	ʈ^{h}		k^{h}	
b	d		dʒ	ɖ		g	
		s	ʃ				h
		z					
m	n						
	r						
	l						
w					j		

Description

Kachru writes that

> /ɖ/ and [ɽ] are intervocally and finally in free variation. By the large, [ɽ] is used by both Muslims and Hindus in village Kashmiri, while in Srinagar it is found only in the speech of some Muslims. (p.60)

Discussion

The retroflex flap only exists in Kashmiri as a result of the flapping of ɖ, and therefore cannot be confused with other phonemes in the language. Hence this process does not neutralize any contrasts.

Notes on Difference from Kirchner's list

Kirchner (1998) lists the lenition process in Kashmiri as involving the alternation ʈ, ɖ - ɾ, suggesting that there is a neutralization between the voiced and voiceless retroflex stops. I follow his grammar source (Kachru 1960) more closely in describing the process as an alternation between the voiced retroflex stop and the retroflex flap (ɖ - ɽ). Kachru never mentions the voiceless retroflex as undergoing a flapping process, or the possibility that flapping involves an additional change in place of articulation.

3.57. Kirghiz

Language:	Kirghiz	*Dialect*:	Not specified
Phylum:	Altaic, Turkic, Western, Aralo-Caspian		
Inventory:	Kirchner	*Source*:	Hebert & Poppe 1963
Lenition(s):	β → w / V_V		
Type:	Gliding		
Classification:	Not Neutralization		

Consonant Inventory

Table 66: Phonemic Consonant Inventory of Kirghiz

p		t	ts	tʃ		k
b		d		dʒ		g
	f		s	ʃ		x
			z			
m		n				ŋ
		r				
		l				
w					j	

Description

Describing the phoneme w, Hebert and Poppe write that it is "a bilabial semivowel intervocalically, as in English cow; a voiced bilabial fricative elsewhere" (p.3).

Discussion

There are no other details on the alternation β - w in the source, but the inventory of phonemes and the complementary distribution between β and w are enough to determine that no contrasts are obliterated by the gliding of β: this sound only exists in the language as a result of the gliding process.

3.58. Korean

Language:	Korean	*Dialect*:	Not specified
Phylum:	Language Isolate		
Inventory:	Kirchner	*Source*:	Martin 1992
Lenition(s):	Stops voiced between voiced sounds		
Type:	Voicing		
Classification:	Not Neutralization		
Difference:	Two processes omitted		

Consonant Inventory

Table 67: Phonemic Consonant Inventory of Korean

p	t	ts		k	
p:	t:	ts:		k:	
p^h	t^h	ts^h		k^h	
		s			h
		s:			
m	n			ŋ	
	l				
w			j		

Description

Martin reports that in Korean "between typically voiced sounds (vowels or semivowels, nasals, the liquid), the lax consonants are lightly voiced in rapid speech" (p.27).

(3.50) Example of voicing in Korean (p.27):

a:nta [a:nda] 'knows'

Discussion

According to the phoneme inventory for Korean, voicing is not contrastive. Therefore the voiced counterparts surface in the language only as results of this voicing process and do not neutralize with existing phonemes.

Notes on Difference from Kirchner's list

Kirchner (1998) lists three lenition processes for Korean, adding to the voicing one included here the loss of h between two voiced sounds and the loss of w before a non-high vowel. The description in Kirchner's language source (Martin 1992) however suggests that these processes are not synchronic alternations, or any type of phonological processes, but instead mere spelling conventions. For example, about the elided h Martin writes:

> From a strictly synchronic point of view, there is no h in words like man(h)i, despite the spelling, which is morphophonemic; we infer the h from other forms such as mankho ← manh.ko, as it is spelled. In unfamiliar Chinese words, the underlying h will emerge as a kind of reading pronunciation, but it usually drops when the word comes to be often said. (p.35)

Hence there is no reason to believe that such h's are part of the grammar of Korean, but instead, are only relics of the spelling. Speakers of Korean have also assured me that there is no dropping of w. I therefore omit both these processes from my corpus.

3.59. Kuna (Paya)

Language:	Kuna[12]	*Dialect*:	Paya
Phylum:	Chibchan, Kuna		
Inventory:	Lavoie	*Source*:	Pike, Forster & Forster 1986
Lenition(s):	b d g → β ð ɣ / V_V (following a stressed syllable)		
Type:	Spirantization		
Classification:	Not Neutralization		

[12] Most likely Kuna and Cuna (§3.21) refer to the same language.

Trends: An additional example of a sound merger without neutralization in the language

Consonant Inventory

Table 68: Consonant Inventory of Paya Kuna

p	t		tʃ		k	
	tʷ		tʃʷ		kʷ	
b	d		dʒ		g	
	dʷ		dʒʷ		gʷ	
		s				h
		z				
m	n					
mː	nː					
mʷ						
	ɾ					
	r					
	l					
	lː					
w				j		

The authors include the labialized stops in their inventory but note that these are rare.

Description

In the Paya dialect of Kuna b d g are voiced fricatives when word medial, intervocalic, and following a stressed vowel.

(3.51) Examples of intervocalic spirantization (p.459):

a. paβa 'father'
b. peðe 'you'
c. naɣa 'foot'

Discussion

The voiced fricatives only surface in the language as results of the spirantization process in question and therefore do not neutralize with existing phonemes.

Additional Comments

This language exhibits an example of phonetic neutralization that does not lead to phonological neutralization. Intervocalically there is a loss of the lenis/fortis distinction in consonants, but the contrast is maintained by shifting to vowel length: the lenis phonemes are preceded by an allophonically lengthened vowel while the fortis ones are preceded by a shortened vowel. I do not add the surface change between lenis and fortis to the inventory of processes because this phenomenon is not described very clearly in the source, especially in terms of how the lenis/fortis distinction is manifested. In general, it appears that for obstruents the lenis fortis distinction is manifested in voicing, and for nasals and laterals in length.

3.60. Kupia (Sujanakota)

Language: Kupia *Dialect*: Sujanakota Village in Manchinaput Block, Visakhapatnam, Andhra Pradesh
Phylum: Indo-European, Indo-Iranian, Indo-Aryan, Eastern zone, Oriya
Inventory: Kirchner *Source*: Christmas & Christmas 1975
Lenition(s): p ʈ ɖ → ɸ ŧ ɽ / V_V in unstressed position (and _C for ɖ)
Type: Spirantization
Classification: Not Neutralization
Trends: Flapping patterns with spirantization
Difference: Sounds participating in the process are corrected

Consonant Inventory

Table 69: Phonemic Consonant Inventory of Sujanakota Kupia

p	t		tʃ	ʈ		k
b	d		dʒ	ɖ		g
		s				
m	n					ŋ
	ɾ					
	l					
w					j	

Description

Intervocalically in unstressed position p freely varies with ɸ (3.52a, b), and ʈ freely varies with ŧ, which the authors describe as a voiceless retroflex with a grooved pronunciation (3.52c). ɖ is a retroflex flap ɽ in all its intervocalic occurrences (3.53).

(3.52) Intervocalically in unstressed positions (p.18):

a.	/kuːpija/	['kuːpiʲa] ~ ['kuːɸiʲa]	'Kupia'
b.	/suːpu/	['suːpu] ~ ['suːɸu]	'winnowing basket'
c.	/tʃaːʈu/	['tsaːʈu] ~ ['tsaːŧu]	'leg'

(3.53) Intervocalic occurrences of ɖ (p.18):

a.	/ɖeːɖi/	['ɖeːɽi]	'pillar'
b.	/goːɖo/	['goːɽo]	'horse'
c.	/laːɖu/	['laːɽu]	'spittle'

ɖ is unflapped preceding n, r, l (3.54a, b) but before any other consonant it is flapped (3.54c, d).

(3.54) ɖ before consonants (p.22):

a.	/waɖɳi/	[waɖɳi]	'broom'
b.	/dʒoːɖɭu/	[dzoːɖɭu]	'shoes'
c.	/taɖpa/	[ʈ̥aɽpa]	'itch'
d.	/paɖto/	[paɽ̥to]	'behind'

Discussion

None of the results of the spirantizing/flapping process described here (ɸ ŧ ɽ) are part of the phoneme inventory for Kupia. Hence no contrast is obliterated by these alternations as every occurrence of these sounds can only be interpreted as the phonemes p ʈ ɖ.

Notes on Difference from Kirchner's list

Kirchner (1998) lists the spirantization/flapping for Kupia as involving the following alternations: p - ɸ, ʈ - r, ɖ - r. If this were true, not only would the flapping of the retroflex stops neutralize their voicing contrast, but it would also lead to neutralization between the retroflex stops and the existing flap phoneme in the language. I follow Kirchner's grammar source (Christmas & Christmas 1975) in describing the alternations as not changing the place of articulation, or the voicing contrast, of the affected sounds. The authors are very clear about the fact that the retroflex sounds, in their lenited form, remain retroflex, and that the distinction between the flapped ʈ and ɖ (to ƫ and ɽ) is maintained.

3.61. Ladakhi (Central)

Language:	Ladakhi	*Dialect*:	Central Ladakhi (standard)
Phylum:	Sino-Tibetan, Tibeto-Burman, Himalayish, Tibeto-Kanauri, Tibetic, Tibetan, Western, Ladakhi		
Inventory:	Kirchner	*Source*:	Koshal 1976
Lenition(s):	b d g - β ð ɣ		
Type:	Spirantization		
Classification:	Not Neutralization		

Consonant Inventory

Table 70: Phonemic Consonant Inventory of Central Ladakhi

p	t	ts	tʃ	ʈ		k	
pʰ	tʰ	tsʰ	tʃʰ	ʈʰ		kʰ	
b	d	dz	dʒ	ɖ		g	
		s	ʃ	ʂ			h
		z	ʒ				
m	n				ɲ	ŋ	
	r						
	l						
	ɬ						
w					j		

Description

In Ladakhi b d g in initial clusters after r, l and after a vowel or r, l, n, ŋ and a syllable boundary are spirantized to β ð ɣ. The bilabial and alveolar are in free variation with their spirant counterparts in these contexts, but the velar undergoes obligatory spirantization.

(3.55) b ~ β (p.28):

a.	[rβət]	'an instrument to smooth the harrow'
b.	[nʌr-βu]	'gem'
c.	[gər-βa]	'black smith'
d.	[miŋ-βo]	'brother'

(3.56) d ~ ð (p.30):

a.	[rðɛ-mo]	'beautiful'
b.	[rðuŋ-ma]	'rafter'
c.	[rin-ði]	'kind of metal'

(3.57) g - ɣ (p.34):

a.	[rɣəd-mo]	'old woman'
b.	[lə-ɣur]	'immediate'
c.	[sɛr-ɣər]	'goldsmith'

Discussion

The inventory of phonemes for Ladakhi shows that spirantization is not contrastive for obstruents. The stops b d g and their spirant counterparts are listed as allophones of each other in the source, and are also in complementary distribution. The spirantization of b d g does not neutralize any contrasts because β ð ɣ obviously do not represent any phonemes except these stops.

3.62. Lahaul (Pattani)

Language:	Lahaul	*Dialect*:	Pattani
Phylum:	Sino-Tibetan, Tibeto-Burman, Himalayish, Tibeto-Kanauri, Western Himalayish, Kanauri		
Inventory:	Kirchner	*Source*:	Sharma 1982
Lenition(s):	1. p^h ~ p 2. b d g → b̥ d̥ g / _[-voi]C, _# 3. ɖ → ɻ / V_V, _# 4. h → weak h / V_V 5. Pre-pausal stops are unreleased (rare) 6. b d g → $b^ʔ$ $d^ʔ$ $g^ʔ$ in coda (rare)		
Type:	1. Deaspiration 2. Devoicing 3. Gliding 4. Deaspiration 5. Unrelease 6. Glottalization		
Classification:	1. Limited Neutralization 2. Not Neutralization 3. Not Neutralization 4. Not Neutralization 5. Not Neutralization		

6. Not Neutralization

Difference: Description of process 3 is corrected

Consonant Inventory

Table 71: Consonant Inventory of Pattani Lahaul

p	t		tʃ	ʈ	c	k	(ʔ)
p^h	t^h		$tʃ^h$	$ʈ^h$	c^h	k^h	
b	d		dʒ	ɖ	ɟ	g	
(b^h)	(d^h)		($dʒ^h$)	($ɖ^h$)	($ɟ^h$)	(g^h)	
		s		ʂ	ç		h
		z					
m	n			ɳ	ɲ	ŋ	
m^h	n^h						
				ɽ			
				$ɽ^h$			
	l						
	l^h						
w				ɻ	j		
				$ɻ^h$			

The sounds in parentheses are definitely not phonemic; also, the palatal plosives do not contrast (that is, c c^h ɟ are members of the same phoneme). Other sounds may not be phonemic as well.

Description

Process 1 concerns the aspiration of p^h: in comparison to its occurrence in accented syllables (e.g., p^hukə 'body'), in pre-accented syllables the amount of aspiration is reduced (e.g., p^hukan 'flour'). Medially and finally p^h is in free variation with p (3.58).

(3.58) p^h ~ p (p.48):

a. ʤəgegp^{h}i ~ ʤəgegpi 'to tremble/to shiver'

b. hjup^{h}tʃi ~ hjuptʃi 'to open'

Process 2 is the reduction in the voicing of b d g preceding a voiceless consonant and word-finally. The author does not include this aspect of the voiced stops in his transcription. Process 3 is the intervocalic and word-final gliding of the retroflex voiced stop. Process 4 is the intervocalic weakening of the aspiration of h. Process 5 is the fact that pre-pausal stops in the language are unreleased, and process 6 is the glottalization of voiced stops in coda position, which is more likely an example of fortition than lenition.

Discussion

Process 1, the deaspiration of p^h, occurs in three environments: in pre-stressed syllables, word-medially, and word-finally. In unstressed syllables the aspiration is less copious than in stressed ones but not altogether missing, so there is no neutralization between p^h and p. Word-finally p is rare, and p^h is not attested in any native words but instead is limited to rare occurrences in loan words. Therefore, in this context neutralization between p^h and p is also avoided. Word-medially the aspirated and simple stops are in free variation, but phonetically there is indeed neutralization between the two. Since phonetic neutralization results in only one of the 3 positions in which this process takes place, and since in that position the alternation is optional, actual phonological neutralization from deaspiration of p^h is quite limited.

Process 2, the reduction in voicing for b d g before a voiceless consonant is not neutralizing because voicing is reduced but not completely missing, which suggests contrast with the voiceless series is maintained. Word-finally, not only are b d g rare, but when they are found there they tend to glottalize (process 6), so neutralization with p t k is surely avoided. It is not completely clear how this process and process 6 interact, but they seem to affect the same sounds in the same environments.

Process 3, the alternation ɖ - ɻ is also not neutralizing: the glide is described as an allophone of the stop, appearing in the language only where it represents the ɖ phoneme. Furthermore, the glide is in complementary distribution with the ɖ, so the two never appear in the same context in the language.

Process 4, like the 2nd process, is a reduction rather than the complete disappearance of some feature of the pronunciation of a sound. This reduction in the strength of aspiration of h intervocalically does not lead to neutralization between h and other sounds in the language.

Process 5 is the lack of plosion in pre-pausal stops, but it rarely applies because pre-pausal stops in Pattani are extremely uncommon. The unrelease of stops, in the few cases where it applies, does not obliterate contrasts in the language because none depend on release for their contrast. Likewise, process 6 does not obliterate contrasts in the language because glottalization only surfaces in Pattani as a result of this rule. Furthermore, as with the previous process, glottalization should rarely apply since stops in coda position are rare.

Additional Comments

In general, plosives in Pattani regularly appear and are contrastive word-initially, show some medial weakening and even a limited amount of neutralization in this context, and are almost nonexistent word-finally. While one cannot draw too many conclusions from this observation without further investigation, I would venture to theorize that the functional load of non-initial plosives in the language is probably rather low.

Notes on Difference from Kirchner's list

Kirchner (1998) lists as process 3 the alternation between ɖ and ɽ. His grammar source (Sharma 1982), however, specifies that the ɖ alternates with a retroflex glide ɻ, not the flap. Hence in this corpus I describe the gliding, rather than the flapping, of ɖ.

3.63. Lama

Language: Lama (aka. Lamba) *Dialect*: Not specified
Phylum: Niger-Congo, Atlantic-Congo, Volta-Congo, North, Gur, Central, Southern, Grusi, Eastern
Inventory: Lavoie *Source*: Ourso & Ulrich 1990
Add. Source: Manessy 1969 (unavailable, but cited in Ourso & Ulrich 1990)
Lenition(s): p → w word-finally
Type: Spirantization/Gliding
Classification: Neutralizing

Consonant Inventory

Table 72: Phonemic Consonant Inventory of Lama (Ourso & Ulrich 1990)

p		t		c	k	kp	
	f		s				h
m		n		ɲ			
		r					
		l					
w				j			

Description

Word initially and intervocalically p and w contrast (3.59) but in syllable codas, where only sonorants can occur in Lama, p and w alternate (3.60). After a homorganic nasal of the demonstrative w and p alternate as well (3.61).

(3.59) p/w contrast (p.136):

a.	wetu	'to sell'
b.	pelu	'to cut'
c.	awor	'place, position'
d.	apeer	(proper name)

(3.60) p - w alternation (p.136):

a.	kpap-ə	'to be similar'
b.	kpaw-s-u	'to reconcile'
c.	jap-ə	'to buy'
d.	jaw	'buy!'

(3.61) w - p alternation (p.136):

a.	wa	'they (class 2)'
b.	mpa	'those (class 2)'

Discussion

The contrast between p and w is neutralized in syllable coda so the spirantization/gliding of the bilabial obviously leads to neutralization. The extent of this neutralization is difficult to assess. Interestingly, the Lama phoneme w is derived from *b in Proto-Gurunsi, and corresponds to the voiced alveolar in related languages (Manessy 1969, cited in Ourso & Ulrich 1990).

3.64. Lamani

Language:	Lamani *Dialect*: Not specified
Phylum:	Indo-European, Indo-Iranian, Indo-Aryan, Central zone, Rajasthani, Unclassified
Inventory:	Kirchner *Source*: Trail 1970
Lenition(s):	ɖ → ɽ except initially or after a homorganic nasal or lateral
Type:	Flapping
Classification:	Not Neutralization
Difference:	Sounds participating in the alternation are corrected

Consonant Inventory

Table 73: Phonemic Consonant Inventory of Lamani

p	t	ts	ʈ		k	
b	d	dz	ɖ		g	
		s				h
m	n		ɳ		ŋ	
	r					
	l		ɭ			
w				j		

Description

Trail writes that

> /ɖ/ has two allophones: intervocalically, in consonant clusters with non-homorganic consonants and /ʈ/, and word finally it is a voiced retroflex post alveolar flap [ɽ]. Elsewhere it is a voiced retroflex post alveolar stop. (p.11)

Discussion

The retroflex flap ɽ only surfaces in the language as a result of the flapping of ɖ so this process does not obliterate distinctions with other phonemes and is therefore not neutralizing.

Notes on Difference from Kirchner's list

Kirchner lists the alternation in Lamani as ɖ - ɾ, but I follow his grammar source (Trail 1970) more closely in describing the result of the flapping of the retroflex stop as a retroflex flap.

3.65. Laotian (Vientiane)

Language:	Laotian	*Dialect*:	Spoken in Vientiane
Phylum:	Tai-Kadai, Kam-Tai, Be-Tai, Tai-Sek, Tai, Southwestern, East Central, Lao-Phutai		
Inventory:	Kirchner	*Source*:	Morev et al. 1979
Lenition(s):	1. Stops are unreleased ('implosive') in final position 2. w j → u i in coda 3. Aspirates and voiced stops prohibited except in onset		

Type: 1. Stop release
2. Vocalization
3. Distributional property

Classification: 1. Not Neutralization
2. Not Neutralization
3. Not Neutralization

Consonant Inventory

Table 74: Consonant Inventory of Vientiane Laotian

p		t		c	k	ʔ
p^{h}		t^{h}			k^{h}	
				c^{w}	k^{w}	$ʔ^{w}$
		t^{hw}			k^{hw}	
b		d				
	f		s			h
			s^{w}			
m		n		ɲ	ŋ	
					$ŋ^{w}$	
		l				
		l^{w}				
w				j		

Description

Process 1 is that "in the initial position, the stops /p t k ʔ/ are realized as the explosives [p t k ʔ], and in the final position as implosives [p˺ t˺ k˺ ʔ˺]" (p.20).

Process 2 is that "the phoneme /w/ is realized as [u̯] in the final position and as a fricative sonant [w] in the initial position. The phoneme /j/ is realized as /i/ at the end of a syllable, and as /j/ at the beginning of a syllable" (p.20).

Process 3 is that

> [T]he consonants /p t k ʔ m n ŋ w j/ may occur in syllables either in the prevocalic or in the postvocalic position. All the other consonants occur only at the beginning of a syllable. (p.20)

Discussion

Process 1, the unrelease of voiceless obstruents in final position, is not neutralizing because the implosive/explosive quality of these stops is not contrastive. Process 3 is also not neutralizing because it refers to a distributional property of the language and does not involve an active process, although it does suggest that there might have been some active process that led to this situation in the past.

Process 2, the alternation between the glides and their corresponding vowels, is a more complicated story. The w alternates with a non-syllabic vowel u, which is not a phonemic vowel in the language; this alternation, therefore, does not neutralize existing contrasts. The j, on the other hand, alternates with a long i, which is an allophone of the phonemic vowel i in Vientiane Laotian so there is potential for neutralization.

Certain properties of Lao syllables come into play in this situation. The authors divide Lao syllables into several categories: "open (syllables without a final consonant), half-open (with a final /j/ or /w/), half-closed (with a final /m/, /n/ or /ŋ/) and closed (syllables with a final stop consonant /p/, /t/, /k/ or /ʔ/)" (p.21). For half-open syllables, the ones of interest to us, the choice of a labial (w) or non-labial glide (j) is predictable: front vowels i iə e ɛ are followed by w and back vowels u uə ɯ ɯə ɔ by j; after o and ɤ both endings are possible.

Since the coda glide of a half-open syllable is predictable the vocalization of j after one of the 5 back vowels makes the presence of the j phoneme obvious (except in the limited case of the vowels that act as exception to the half-open rule, o ɤ). And while open syllables in the language always end with the long allophones of vowels, the vowel clusters resulting from u uə ɯ ɯə ɔ followed by long i are not found elsewhere in Laotian.[13] In other words, syllable final ui uəi ɯi ɯəi ɔi only surface in the language as a result of the j - i alternation, and therefore this alternation does not neutralize any contrasts.

3.66. Lezgian

Language:	Lezgian	*Dialect*:	Standard
Phylum:	North Caucasian, Northeast, Lezgian		
Inventory:	Lavoie	*Source*:	Haspelmath 1993
Lenition(s):	p k t ts tʃ q → b g d z ʒ ʁ / _#		
Type:	Voicing		
Classification:	Not Neutralization		
Trends:	Phonetic, but not phonological, neutralization		
Difference:	Sounds participating in the process are corrected		

[13] The only attested vowel clusters are diphthongs that end in ə, double/long vowels, and a very small number of words with a vowel + a combination.

Consonant Inventory

Table 75: Consonant Inventory of Lezgian

p		t	ts	tʃ		k	q	ʔ
p^{h}		t^{h}	ts^{h}	tʃ^{h}		k^{h}	q^{h}	
p'		t'	ts'	tʃ'		k'	q'	
		t^{w}	ts^{w}			k^{w}	q^{w}	
		t^{hw}	ts^{hw}			k^{hw}	q^{hw}	
		t'^{w}	ts'^{w}			k'^{w}	q'^{w}	
b		d				g		
						g^{w}		
	f		s	ʃ		x	χ	h
			s^{w}				χ^{w}	
			z	ʒ			ʁ	
			z^{w}				ʁ^{w}	
m		n						
		r						
		l						
w					j			

Description

Under the section for "Phonological Alternations" Haspelmath writes that

> In all monosyllabic nouns with a root-final unaspirated voiceless stop, this stop alternates with the corresponding voiced stop in word-final position (i.e. in the Absolutive singular form). This automatic alternation is motivated by the ban on word-final unaspirated stops. (p.53)

Discussion

Haspelmath does not make it clear whether his inventory of consonants in the language is phonemic, so classifying the word-final process he describes as neutralizing or not is not so straightforward. However, the author does specify that this alternation is limited to a handful of nouns, providing a "fairly complete list of nouns exhibiting word-final unaspirated voicing" of 47 items. Considering the fact that 6 alternations (p - b, k - g, t - d, q - ʁ, ts - z, tʃ - ʒ) are spread out over only 47 lexical items, the number of affected nouns per alternation is very small. Since the actual number of affected lexical items is small, the chances of the alternations in question leading to the obliteration of meaning distinctions are also very small. Hence, due to low functional load, the word-final voicing of stops in Lezgian is considered not neutralizing.

Additional Comments

This voicing of stops word-finally is probably an example of fortition rather than lenition.

Notes on Difference from Kirchner's list

Lavoie (1996), as cited in Kirchner (1998), lists the lenition in Lezgian as "pV → b / _#". I follow Lavoie's grammar source (Haspelmath 1993) in describing the process as affecting the entire unaspirated stop/affricate series.

3.67. Limbu (Phedāppe)

Language:	Limbu	*Dialect*:	Phedāppe dialect of Yakthuŋba[14]
Phylum:	Sino-Tibetan, Tibeto-Burman, Himalayish, Mahakiranti, Kiranti, Eastern		
Inventory:	Kirchner	*Source*:	van Driem 1987
Lenition(s):	1. p tʃ k → b dʒ g / [nasal]_, ʔ_, V_V 2. p^h t^h k^h → b^h d^h g^h / V_V 3. t → ʔl ($ʔt^l$) / _# or _{ʔ, h} 4. b → w / V_V or after (non-homorganic) nasal (opt.)		
Type:	1. Voicing 2. Voicing 3. Lateralization /debuccalization 4. Gliding		
Classification:	1. Not Neutralization 2. Not Neutralization 3. Not Neutralization 4. Limited Neutralization		
Trends:	Phonemic overlap between processes 1 and 4 (also between process 3 and another process in the language)		

[14] Limbu dialect spoken in Tamphulā village, Tehrathum district in Eastern Nepal's Kośī zone.

Consonant Inventory

Table 76: Phonemic Native Consonant Inventory of Phedāppe Limbu

p	t		tʃ		k	ʔ
p^h	t^h				k^h	
(b)						
		s				h
m	n				ŋ	
	r					
	l					
w				j		

Description

Following a nasal or a glottal stop and in intervocalic position p tʃ k voice to b dʒ g (process 1). The aspirated sounds p^h t^h k^h voice intervocalically, and in the case of the bilabial and the alveolar also after a nasal or a glottal stop, to $b^ɦ$ $d^ɦ$ $g^ɦ$ (process 2). Word finally and before ʔ or h the alveolar stop t is realized as ʔl (process 3). b is a marginal phoneme in Limbu as it appears in only a handful of native words, but where it does, it optionally glides to w in intervocalic or postnasal, as long as it isn't m, position (process 4).

Discussion

The first two processes involve voicing of stops in similar, but not exactly identical, contexts. With the exception of the marginal phoneme b, voicing is not contrastive in Limbu. Voicing of stops results in sounds that do not already exist in the language, and therefore, is not neutralizing.

The lateralization of the alveolar stop t results in a sound that the author describes as $ʔt^l$ or ʔl. The lateral aspect of this sound is not complete, but regardless, neutralization with an existing lateral in the language is still avoided because in the same environment the phonemic lateral surfaces as a trill. Hence process 3 is also not neutralizing.

The 4th process is the gliding of b to w, and it occurs in a similar context to where p is voiced to b. w does appear to be phonemic in Limbu, but b is a marginal phoneme and gliding is optional so neutralization of meaning distinctions, if it occurs at all, is probably minimal. For this reason I classify the gliding of b as "limited neutralization".

Additional Comments

The case of p - b and b - w in a similar environment is an example of phonemic overlap. Although b is only a marginal phoneme in Limbu, neutralization with the voiceless phoneme is still avoided by gliding the b in the same context where p is voiced. Process 3 is also part of a phonemic overlap where the existing lateral in the language alternates with a trill in the same environment where the alveolar surfaces as a lateral.

3.68. Lotha

Language:	Lotha	*Dialect*:	Not specified
Phylum:	Sino-Tibetan, Tibeto-Burman, Kuki-Chin-Naga, Naga, Ao		
Inventory:	Kirchner	*Source*:	Acharya 1983
Lenition(s):	1. t → d / n_		
	2. k → g / V_V		
Type:	1. Voicing		
	2. Voicing		
Classification:	1. Not Neutralization		
	2. Not Neutralization		

Consonant Inventory

Table 77: Phonemic Consonant Inventory of Lotha

p	pf	t	ts	tʃ		k	
p^h		t^h	ts^h	tʃ^h		k^h	
	bv						
	f		s	ʃ			h
	v		z	ʒ			
m		n			ɲ	ŋ	
m^h		n^h			ɲ^h	ŋ^h	
		r					
		r^h					
		l					
		l^h					
w					j		

Description

In Lotha t voices medially when it follows a homorganic nasal (3.62) and nowhere else (3.63).

(3.62) Examples of t voicing (p.32):

a.	/mʰo˩˥nta˩˥/	[mʰo˩˥nda˩˥]	'treaty'
b.	/vo˧nto˥n/	[vo˧ndo˥n]	'hungry'
c.	/wo˥nta˥ŋ/	[wo˥nda˥ŋ]	'light'

(3.63) t not voiced intervocalically (p.33):

a.	/ta˧tʰi˩˥/	[ta˧tʰi˩˥]	'nature'
b.	/ti˧ta˩˥/	[ti˧ta˩˥]	'peacock'
c.	/o˧to˥n/	[o˧to˥n]	'forehead'

k is voiced in intervocalic position (3.64) but nowhere else (3.65).

(3.64) Examples of k voicing intervocalically (p.34)

a.	/sə˧nrə˧ka˩˥/	[sə˧nrə˧ga˩˥]	'rainbow'
b.	/te˧krə˩˥/	[te˧grə˩˥]	'honey bee'

(3.65) k is not voiced elsewhere (p.35):

a.	/jo˥ŋko˩˥k/	[jo˥ŋko˩˥k˺]	'spoon'
b.	/ku˥rə˩˥/	[ku˥rə˩˥]	'horse'

Discussion

Voicing is contrastive for fricatives and affricates in Lotha, but not for stops. Hence d and g only surface in the language as results of the voicing processes in question and do not neutralize with existing phonemes.

Additional Comments

Voicing in Lotha does not affect the entire stop series and it does not apply in the same environment for t and k. For this reason the two voicing processes in Lotha are listed as separate, as they are in Kirchner's inventory.

3.69. Lowland Murut (Timugon: Poros)

Language: Lowland Murut *Dialect*: Poros sub dialect of Timugon dialect

Phylum:	Austronesian, Malayo-Polynesian, Western Malayo-Polynesian, Borneo, Northwest, North Sarawakan, Dayic, Murutic, Northern
Inventory:	Kirchner *Source*: Halle & Clements 1983
Add. Source:	Prentice 1971 (Halle & Clements' language source)
Lenition(s):	b d g → β ɾ ɣ after vowels, semi-consonants and d g ʔ, and clause-finally
Type:	Spirantization
Classification:	Not Neutralization
Trends:	Flapping patterns with spirantization

Consonant Inventory

Table 78: Phonemic Consonant Inventory of Timugon Lowland Murut (Prentice 1971)

p	t				k	ʔ
b	d		dʒ		g	
		s				
m	n				ŋ	
	l					
w				j		

Description

The voiced stops b d g are realized as β ɾ ɣ after vowels, semi-consonants and d g ʔ and clause-finally.

Discussion

According to the phoneme inventory for Lowland Murut spirantization is not contrastive. Thus, the outcome of the spirantization process cannot be confused with existing sounds in the language and therefore does not lead to neutralization.

3.70. Macushi

Language:	Macushi *Dialect*: Not specified
Phylum:	Carib, Northern, East-West Guiana, Macushi-Kapon, Macushi
Inventory:	Lavoie *Source*: Abbott 1991
Lenition(s):	p t s k → b d z g following long syllables
Type:	Voicing

Classification: Not Neutralization

Consonant Inventory

Table 79: Phonemic Consonant Inventory of Macushi

p	t			k	
		s			h
m	n				
	ɾ				
w			j		

The glottal h occurs only in syllable-final position.

Description

The stops p t k have voiced variants that occur "following a 'long' syllable (CVV, CVC or VC) in which there is a long vowel or a final n or h (this does not include all cases of rhythmically long syllables)" (p.140). In (3.66-3.68) the voiced variants are shown in (a) and the voiceless in (b).

(3.66) Allophones of p (p.140):

a.	paapa	[paaaba]	'father'
	iipɨ	[iibɨ]	'come'
	sunpa	[sumba]	'tray, shallow basket'
	ahpu	[ahbu]	'your foot'
b.	pɨrɨu	[pɨrɨu]	'arrow'
	kapoi	[kapoi]	'moon'

(3.67) Allophones of t (p.141):

a.	jepotorɨ	[ðepoodorɨ]	'lord'
	unta	[unda]	'my mouth'
	mohta	[mohda]	'move'
b.	pata	[pata]	'place'

(3.68) Allophones of k (p.141):

a.	paaka	[paaga]	'cow'
	ahka	[ahga]	'light'
	inkamoro	[ingamoro]	'those'
b.	kaane	[kaane]	'no'
	iwarɨka	[iwarɨka]	'monkey'

The spirant s is voiced in the same environment as the stops (3.69a) except when the preceding or following vowel is a high front vowel where the ʒ variant occurs (3.69b). Elsewhere in the vicinity of high vowels s surfaces as ʃ (3.69c) and everywhere else as s (3.69d).

(3.69) Allophones of s (P.141):

a.	aase	[aaze]	'let's walk'
	ujaponse	[ujaboŋze]	'my bench'
	juhse	[juhze]	'want'
b.	insenan	[iŋʒenaŋ]	'these'
	ujensi	[ujeŋʒi]	'my daughter'
	tiwinserɨ	[tiwiŋʒerɨ]	'alone'
c.	sunpa	[ʃumba]	'tray'
	sikɨ	[ʃikɨ]	'flea'
	isan	[iʃaŋ]	'his mother'
d.	sahman	[sahmaŋ]	'hard'

Discussion

The inventory of phonemes in Macushi makes it clear that voicing is not contrastive for stops or the spirant, and therefore, the results of the voicing process do not neutralize with existing sounds in the language.

3.71. Maidu

Language:	Maidu	*Dialect*:	Not specified
Phylum:	Penutian, Maiduan		
Inventory:	Lavoie	*Source*:	Shipley 1963
Lenition(s):	p t k → $p^ʔ$ $t^ʔ$ $k^ʔ$ before a consonant or a juncture		
Type:	Glottalization		
Classification:	Not Neutralization		
Difference:	Restated the process		

Consonant Inventory

Table 80: Phonemic Consonant Inventory of Maidu

p	t		tʃ		k	ʔ
p^h	t^h		$tʃ^h$		k^h	
b	d					
		s				h
m	n					
	l					
w				j		

Description

The author describes the alternation between simple and glottal stops as:

> Any morphophonemic glottalized stop, in the environment before a consonant, or a juncture, is represented phonemically by the corresponding unglottalized voiceless stop, e.g., morphophonemic ||tep^{ʔ}ni.|| is phonemic /tepni./ 'with a banded hand-game bone'. Similarly morphophonemic ||tep^ʔ.|| is phonemic /tep./ (call used for choosing the banded hand-game bone). (p.2)

Discussion

Shipley reports that Maidu has two phonemically voiceless stop series: one that is aspirated, and one that includes the simple stops (p t k) and the glottalized stops ($p^ʔ$ $t^ʔ$ $k^ʔ$) as its allophones. Regardless of whether the simple or the glottalized stops are considered the basic phonemes (I list the simple stops in the inventory above) it is clear that the alternation between them does not neutralize contrast with other phonemes in the language.

Notes on Difference from Kirchner's list

Lavoie (1996), as cited in Kirchner (1998), lists "deglottalize ejective stops finally" as a lenition process in Maidu. Her grammar source (Shipley 1963) however does not include ejectives as part of the language's inventory of sounds. There may have been a misunderstanding on Lavoie's part of Shipley's use of diacritics, but the author calls these stops glottalized rather than ejective. I follow Lavoie's grammar source for the language more closely by describing the process as an alternation between the simple and glottalized stops.

3.72. Malayalam

Language:	Malayalam	*Dialect*:	Not specified
Phylum:	Dravidian, Southern, Tamil-Kannada, Tamil-Kodagu, Tamil-Malayalam, Malayalam		
Inventory:	Kirchner	*Source*:	Mohanan 1986
Lenition(s):	1. p t̪ ʈ k → b d̪ ɖ g / [+son]_, [-nasal]_V 2. b d̪ ɖ g → β ð̪ ɽ ɣ / [+son]_, [-nasal]_V		
Type:	1. Voicing 2. Spirantization		
Classification:	1. Not Neutralization 2. Not Neutralization		
Trends:	Flapping patterns with spirantization; phonemic overlap between the two processes		
Difference:	Description of the 2nd process is corrected		

Consonant Inventory

Table 81: Consonant Inventory of Malayalam

p	t̪	t			tʃ	ʈ	c	k	
pʰ	t̪ʰ				tʃʰ	ʈʰ	cʰ	kʰ	
b	d̪				dʒ	ɖ	ɟ	g	
bʰ	d̪ʰ				dʒʰ	ɖʰ	ɟʰ	gʰ	
				s	ʃ	ʂ			h
m	n̪	n	nʲ			ɳ	ɲ	ŋ	
		ɾ	ɾʲ						
		l				ɭ			
w						ɻ	j		

Table 82: Phonemic Stops and Nasals of Malayalam

p	t̪		tʃ	ʈ	k
pʰ	t̪ʰ		tʃʰ	ʈʰ	kʰ
b	d̪		dʒ	ɖ	g
bʰ	d̪ʰ		dʒʰ	ɖʰ	gʰ
m		n		ɳ	

Description

Process 1 is the voicing of stops when preceded by a sonorant. After non-nasal sounds voicing applies only when the stop in question is followed by a vowel. Process 2 is the spirantization of voiced stops in the same environment where the voiceless series is voiced (that is, in the same environment where process 1 applies).

Discussion

Neutralization from voicing stops is avoided because in the same environment the already voiced stops in the language are spirantized. Since spirants are not phonemic in Malayalam, the results of process 2 do not neutralize with existing sounds either. This is an example of phonemic overlap, where p → b and b → β in the same context, thus avoiding neutralization between the newly voiced b's and the existing b's in the language.

Notes on Difference from Kirchner's list

Kirchner (1998) reports that the spirantization of the retroflex stop results in an alveolar flap, which would neutralize with the existing flap in the language. I follow Kirchner's grammar source (Mohanan 1986) who reports that the retroflex stop alternates with a retroflex flap, which is not phonemic in Malayalam.

3.73. Manobo (Ata)

Language:	Manobo	*Dialect*:	Ata
Phylum:	Austronesian, Malayo-Polynesian, Western Malayo-Polynesian, Southern Philippine, Manobo, Central, South, Ata-Tigwa		
Inventory:	Lavoie	*Source*:	Morey & Post 1964 (in Reid 1971)
Lenition(s):	b d g - β ð ɣ		
Type:	Spirantization		
Classification:	Not Neutralization		

Consonant Inventory

Table 83: Phonemic Consonant Inventory of Ata Manobo

p	t			k	q	
b	d			g		
		s				h
m	n			ŋ		
	l					
w			j			

Description

The only information on Ata Manobo phonology in the source is the inclusion of β ð ɣ as allophones of b d g in the inventory of sounds. Context for these variations is not provided.

Discussion

The spirantization of b d g is not neutralizing because β ð ɣ are only listed in the sound inventory as allophones of the voiced sounds, and not as phonemes in Manobo. That means all occurrences of the spirants are limited to their representation of the b d g phonemes, and therefore, no contrast with other phonemes is neutralized.

3.74. Maori

Language: Maori *Dialect*: Several are described
Phylum: Austronesian, Malayo-Polynesian, Central-Eastern, Eastern Malayo-Polynesian, Oceanic, Central-Eastern Oceanic, Remote Oceanic, Central Pacific, East Fijian-Polynesian, Polynesian, Nuclear, East, Central, Tahitic
Inventory: Lavoie *Source*: Bauer 1993
Lenition(s): k - x/ç (dialect dependent)
Type: Spirantization
Classification: Not Neutralization

Consonant Inventory

Table 84: Phonemic Consonant Inventory of Maori

p		t	k	
	f			h
m		n	ŋ	
		ɾ		
w				

Description

Bauer (1993: 531) describes the frication of k as:

> /k/ shows a good deal of variation in articulation. It is usually advanced velar before front vowels, and retracted velar before back vowels. In isolated tokens in Western dialect speakers throughout the North Island, and with greater regularity in the Bay of islands and particularly Whangaroa areas of Northland, /k/ is realized as a velar fricative primarily but not exclusively before back vowels, and in a few cases as a palatal fricative before front vowels. Thus the following pronunciations of place-names can be heard:

Kaaea	[xaːɛɐ]	Kerikeri	[çɛɾiçɛɾi]
Hopekako	[hɔpɛxɐxɔ]	Rimariki	[ɾimɐɾiçi]

Discussion

Bauer reports that k can be realized as either a velar fricative x or a palatal fricative ç. Neither of these sounds are listed in the inventory of phonemes which means all occurrences of the two fricatives are limited to representations of the k phoneme, and therefore, do not lead to neutralization with other phonemes in Maori.

3.75. Marathi (Halabi)

Language: Marathi *Dialect*: Halabi
Phylum: Indo-European, Indo-Iranian, Indo-Aryan, Eastern zone, Bengali-Assamese
Inventory: Kirchner *Source*: Singh 1977
Lenition(s): ɖ $ɖ^h$ - ɽ $ɽ^h$
Type: Flapping

Classification: Not Neutralization
Difference: The sounds participating in the process are corrected

Consonant Inventory

Table 85: Phonemic Consonant Inventory of Halabi Marathi

p	t		ʈ	c	k	
p^h	t^h		ʈ^h	c^h	k^h	
b	d		ɖ	ɟ	g	
b^h	d^h		ɖ^h	ɟ^h	g^h	
		s				h
m	n				ŋ	
	ɾ					
	l					
w				j		

Description

The retroflex voiced stops in Halabi have retroflex flap allophones as follows: ɖ occurs initially and medially in clusters as the initial member and after homorganic nasal, and ɽ occurs elsewhere; ɖ^h occurs initially, both alone and in clusters, and ɽ^h occurs elsewhere.

Discussion

The flapping of the retroflex voiced sounds does not neutralize any contrasts in the language because the results of this process are not phonemes. The aspiration between ɖ and ɖ^h, which is contrastive in the language, is also preserved (in ɽ and ɽ^h).

Notes on Difference from Kirchner's list

Kirchner (1998) lists the flapping in Halabi as "retroflex stops → ɾ", which on the face of it suggests neutralization among the retroflex stops as well as between them and the existing alveolar flap phoneme. However, his grammar source (Singh 1977) describes the process as affecting only the voiced retroflex stops (ɖ ɖ^h) and alternating with the retroflex flaps (ɽ ɽ^h) as listed above.

3.76. Mataco (Noctenes)

Language: Mataco-Noctenes *Dialect*: Noctenes
Phylum: Mataco-Guaicuru, Mataco
Inventory: Lavoie *Source*: Claesson 1994
Lenition(s): ? h → Ø word-finally, except utterance-finally
Type: Loss
Classification: Not Neutralization
Trends: Phonetic, but not phonological, neutralization
Difference: Description of process is corrected

Consonant Inventory

Table 86: Phonemic Consonant Inventory of Noctenes Mataco

p	t	ts			k^j	q	?
				k^w			
		s		x			h
				x^w			
m	n						
	l						
w			j				

Description

According to Claesson, in the Noctenes dialect of Mataco word-final glottal sounds (? h) in words such as (3.70a, b) are lost utterance-internally, but are retained utterance-finally.

(3.70) Word final glottal sounds (p.25-26):

a.	?atsiinha?	'woman'
	tuunte?	'stone'
	?iihi?	'it is in…'
b.	?ohooyeh	'I go to him/her'
	?ooqoh	'my mother'
	t?uukeh	'(s)he looks for…'

Discussion

The loss of word-final, utterance-internal, glottals constitutes phonetic neutralization both between the two glottals (?/h) and between their existence and absence (?/Ø and h/Ø). But do any of these lead to phonological neutralization? As it turns out, the answer is no: the two glottal sounds do not contrast word-finally so the phonetic neutralization between ? and h does not obliterate meaning distinctions; also, the existence of a glottal sound word-finally is predictable, so its absence does not neutralize with Ø (that is, in the contexts where the glottal sound is phonetically lost, its phonological presence is still obvious and the contrast with Ø is not obliterated).

The possible contrast between the two glottal elements (? and h) is of prime importance to Claesson, and is discussed at length. Claesson rejects other theories that classify the two glottals as members of the same phoneme (e.g., Vinas Urquiza 1970) and shows that word-initially and medially the two contrast (3.71).

(3.71) Word-initial and medial contrast between ? and h (p.35):

a.	?ap	'again'	vs.	hap	'(anaphoric) then, there, that'
b.	?o?is	'I am well'	vs.	?oohis	'my receptacles'
c.	hoo?oh	'hen'	vs.	?ohooho?	'I enter it'

Claesson then argues that the contrast is maintained word-finally: "So, if [?] is a phoneme, why should it not occur as such at the end of words?" (p.35). Despite such a compelling argument as "why shouldn't it", word-finally the glottals do not seem to contrast in Noctenes. Their reported loss in all word-final, phrase-internal, contexts does not constitute neutralization of the contrast between ? and h, because word-finally no such contrast exists.

Claesson reports that no words in Noctenes end with a vowel, lateral, nasal or approximant, but as a general rule such sounds word-finally are followed with a glottal (? or h), which is then lost everywhere but phrase-finally. Hence, since final glottals are predictable in the language, their absence does not lead to neutralization with Ø.

Additional Comments

Claesson's analysis of glottals in Noctenes seems needlessly complicated. The situation itself is rather simple: if a phrase ends with anything but an obstruent it is followed by a glottal. Claesson's interpretation involves an "underlying" glottal at the end of each word that ends with a non-obstruent element, and the subsequent deletion of the glottal in all but phrase-final positions. A much simpler analysis would involve the insertion of a glottal in phrase-final position after non-obstruents. This analysis accounts for the data and is far less complicated for both learners and linguists. Functionally, the proposed analysis does not affect the classification of this process as not neutralizing, but it does suggest the process in question may not be an example of lenition.

Notes on Difference from Kirchner's list

Lavoie (1996), cited in Kirchner (1998), lists the process for Noctenes as "ʔ → h word-finally". Her own language source, however, very clearly and at great length discusses the fact that both ʔ and h are lost word-finally, and never suggests the two alternate.

3.77. Maxakalí

Language:	Maxakalí	*Dialect*:	Not specified
Phylum:	Macro-Ge, Maxakali		
Inventory:	Kirchner	*Source*:	Gudschinsky et al. 1970
Lenition(s):	Stops → vocoids in syllable coda before homorganic C		
Type:	Gliding		
Classification:	Not Neutralization		
Difference:	Description of process is corrected		

Consonant Inventory

Table 87: Phonemic Consonant Inventory of Maxakalí

p	t	c	k	ʔ
				h
m	n	ɲ	ŋ	

Description

> The most unusual characteristic of the Maxakalí phonemic system is the vocalic allophony of the consonants. In general, the basic allophone is replaced by a fully syllabic vocoid, with little or no consonantal closure, in syllable coda before a homorganic consonant (p.82).

The vocalic allophones have the same horizontal tongue position as the consonantal allophones but differ in degree of closure.

Discussion

It is safe to assume that the vocalic allophones of the stops do not neutralize with existing consonants in the language because no such sounds are phonemic. The results of the stop-vocoid alternations also do not neutralize with existing vowels because the vocalic allophones do have some degree of closure and are not fully vowel-like, and more importantly, the language has very strict syllable structure (CVCV) so each sound can easily be distinguished as a consonant or a vowel depending on its position in a syllable. Therefore the gliding, or vocalization, of stops in Maxakalí is not neutralizing.

Notes on Difference from Kirchner's list

Kirchner (1998) describes the alternation in Maxakalí as geminate stops alternating with a glide + singleton stop. I adhere to Kirchner's grammar source (Gudschinsky et al. 1970) in describing the lenition as an alternation between stops and vocoids.

3.78. Mbabaram

Language:	Mbabaram	*Dialect*:	Not specified
Phylum:	Australian, Pama-Nyungan, Paman, Southern Pama		
Inventory:	Kirchner	*Source*:	Dixon 1991
Lenition(s):	Stops voice medially, depending on surrounding sounds		
Type:	Voicing		
Classification:	Not Neutralization		

Consonant Inventory

Table 88: Phonemic Consonant Inventory of Mbabaram

p	t̪	t		c	k
		tʷ			kʷ
m	n̪	n		ɲ	ŋ
		nʷ			
			ɽ		
		r			
		l			
w				j	

Description

Dixon describes voicing in Mbabaram as:

> Voicing appears not to be distinctive for stops. The allophony is as follows: (i) stops are normally voiceless in initial position, in final position, and in medial position after j; (ii) they are voiced after a nasal; (iii) they alternate

between voiced and voiceless in medial position between vowels, or after l, r, or ɽ. (p.355)

Discussion

As Dixon reports both in his prose and list of likely phonemes, voicing is not contrastive for stops in Mbabaram, and the voiced and voiceless variations of stops belong to the same phoneme. Therefore, voicing in this language does not lead to neutralization.

3.79. Middle Chinese

Language:	Middle Chinese	*Dialect*:	More than one discussed
Phylum:	Sino-Tibetan, Chinese		
Inventory:	Lavoie	*Source*:	Pulleyblank 1984
Lenition(s):	χ - x/h		
Type:	Debuccalization		
Classification:	Not Neutralization		
Difference:	Process description is corrected		

Consonant Inventory

Not provided in the source

Description

The only reference to back fricatives discussed in the source is in the table of the progress of initials from Ancient Chinese (AC), to Early Middle Chinese (EMC), to Late Middle Chinese (LMC). The EMC forms in parentheses are Southern.

(3.72) Progress of initials (p.232):

AC	EMC	LMC
χ	x (h)	x

Discussion

Neither x nor h occur elsewhere in the list of initial sound shifts in the source so it is safe to assume that there was no sound merger that resulted from the shift of the χ to x/h, and therefore, this process is not neutralizing.

Notes on Difference from Kirchner's list

Lavoie (1996), cited in Kirchner (1998), lists the process for Middle Chinese as x - h. I follow her language source (Pulleyblank 1984) in listing the shift as χ - x/h, where initial χ shifted to h in the Southern dialects, and x elsewhere.

3.80. Middle Korean

Language:	Middle Korean	*Dialect*: Not specified
Phylum:	Language Isolate	
Inventory:	Lavoie	*Source*: Ramsey 1991
Lenition(s):	b → β / V_V	
Type:	Spirantization	
Classification:	Not Neutralization	

Consonant Inventory

Table 89: Consonant Inventory of Middle Korean

p	t		c	k	ʔ	
pp	tt		cc	kk		
p^h	t^h		c^h	k^h		
		s			h	
		ss			hh	
β		z				
m	n			ŋ		´
	l					´´

The table above is actually a transliteration of the Korean alphabet and not an inventory of sounds in the phonological sense. The author does not explain what he means by the symbols (´ and ´´) in the post-glottal column.

Description

All reference to the spirantization of b in the source is in the proposed reconstruction of *b in Proto-Korean to β in Middle Korean, and an additional note that b → β occurred in intervocalic position.

Discussion

In addition to the fact that β results only from *b and no other proto sounds, the fact that the contrast between *p and *b is preserved is also discussed. The spirantization of b is a diachronic process from which, apparently, no sound merger resulted. Therefore, it could not in itself have led to the neutralization of any contrasts.

3.81. Modern Irish

Language: Modern Irish *Dialect*: Donegal, Connacht, Munster
Phylum: Indo-European, Celtic, Insular, Goidelic
Inventory: Kirchner *Source*: Ó Siadhail 1989
Lenition(s): Dialectal variation in degree of tensing
Type: Dialectal variation
Classification: Not Neutralization

Consonant Inventory

Table 90: Phonemic Consonant Inventory of Modern Irish

p	p^j			t	t^j			k	k^j		
b	b^j			d	d^j			g	g^j		
		f	f^j			s	ʃ			h	h^j
m	m^j			n	n^j						
				r	r^j						
				l	l^j						

The palatal and non-palatal n, r and l are labeled as "tense" in the source.

Description

There is a variation between dialects in the degree of tenseness for the stops that are tense (n n^j r r^j l l^j).

Discussion

The degree of tenseness is not contrastive for either palatalized or non-palatalized n r l, and therefore, a change in this feature does not neutralize any contrasts. Moreover, a dialectal variation does not imply that there is a shift from one dialect to the other, and therefore, does not constitute an active process in the phonology. Hence, neutralization would not have been possible even if tenseness was contrastive.

3.82. Moghamo (Batibo)

Language:	Moghamo *Dialect*: Batibo
Phylum:	Niger-Congo, Atlantic-Congo, Volta-Congo, Benue-Congo, Bantoid, Southern, Wide Grassfields, Narrow Grassfields, Momo
Inventory:	Kirchner *Source*: Stallcup 1978
Lenition(s):	d g → ɾ ɣ / V_V[-stressed]
Type:	Spirantization
Classification:	Not Neutralization
Trends:	Flapping patterns with spirantization; possible phonemic overlap with voicing
Difference:	Sounds participating in the process are corrected

Consonant Inventory

Table 91: Phonemic Consonant Inventory of Batibo Moghamo

p		t	tʃ		k	ʔ
b		d	dʒ		g	
	f		ʃ			
			ʒ			
m		n		ɲ		
w				j		

The author adds that ʔ does not occur initially.

Description

d → ɾ and g → ɣ intervocalically before unstressed vowels. In the case of the velar this is optional.

(3.73) Spirantization of d and g (p.81):

a.	/bedi/	[beɾi]	'to possess'
b.	/sugi/	[sugi] ~ [suɣi]	'to be full'

Discussion

The spirants ɽ and ɣ are not included in the inventory of phonemes, which means they only surface in the language as a result of the process in question and there is no possibility for them to neutralize with existing sounds.

Additional Comments

Stallcup mentions that there is medial voicing of stops, so there appears to be phonemic overlap between voicing of voiceless stops and spirantization of voiced ones in the same context.

Notes on Difference from Kirchner's list

Kirchner (1998) lists the process for Moghamo as "P T K - β ɾ ɣ (context unclear)". I follow Kirchner's grammar source (Stallcup 1978) more closely in my corpus: (a) Stallcup does not give indication that the bilabial sound participates in this alternation; (b) he specifically notes that the alveolar alternates with a retroflex flap, although it isn't clear why that would be the case; and (c) he provides a clear context for this spirantization process.

3.83. Mohawk (Akwesasne)

Language: Mohawk *Dialect*: Akwesasne
Phylum: Iroquoian, Northern Iroquoian, Five Nations, Mohawk-Oneida
Inventory: Kirchner *Source*: Halle & Clements 1983
Add. Source: Bonvillain 1973 (Language source for Halle & Clements)
Lenition(s): t k → d g / _ (w, j)V
Type: Voicing
Classification: Not Neutralization

Consonant Inventory

Table 92: Phonemic Consonant Inventory of Akwesasne Mohawk (Bonvillain 1973)

	t				k	ʔ
			dʒ			
		s				h
	n					
	l					
w				j		

Description

The stops t and k are voiced before vowels. k also voices before a glide (w, j) followed by a vowel, while t never occurs before vowels.

(3.74) Examples of voicing (p.27-28):

a.	/o'liːteʔ/	[o'liːdeʔ]	'pigeon'
b.	/o'tahsa/	[o'dahsa]	'tail'
c.	/'tekeni/	['degeni]	'two'

Discussion

As can be seen from the inventory of phonemes, voicing is not contrastive for stops. The d and g therefore only surface as results of the voicing process and cannot be confused with existing sounds. Hence, the voicing of t and k does not lead to neutralization.

3.84. Mongolian (Kalkha)

Language:	Mongolian	*Dialect*:	Kalkha
Phylum:	Altaic, Mongolian, Eastern, Oirat-Khalkha, Khalkha-Buriat, Mongolian Proper		
Inventory:	Lavoie	*Source*:	Poppe 1970
Lenition(s):	b g (ɢ) - β ɣ (ʁ)		
Type:	Spirantization		
Classification:	Not Neutralization		
Difference:	Process description is corrected, a second process is omitted		

Consonant Inventory

Table 93: Phonemic Consonant Inventory of Kalkha Mongolian

		t	ts	tʃ				
(pʰ)						(kʰ)		
b		d	dz	dʒ		g		
	(f)		s	ʃ			χ	
m		n				ŋ		
		r						
		l						
(w)					j			

Sounds in parentheses occur only in loan words.

Description

The phonemes b and g have numerous allophones depending on their environment. b has the following allophones: b occurs only in medial position, after m or l, and before vowels; β occurs intervocalically, after r in medial position, in syllable-final position before a lax consonant, and in final position in monosyllabic words; b̥ occurs initially, in word-final position in polysyllabic words, and in syllable-final position before a tense consonant; and ɸ occurs in syllable-final position before a tense consonant and in word-final position in words of two or more syllables.

The velar g has allophones which range from velar sounds (g̥ g ɣ) that occur in the vicinity of front vowels to uvular sounds (q ɢ ʁ) that occur only with back vowels, as follows: g and q occur in medial position after m ŋ l r and before a vowel; g̥ and ɢ occur in word-final position, and in syllable-final position before tense consonants; ɣ and ʁ occur in intervocalic position.

Discussion

The fricatives β and ɣ are not phonemic in the language and only surface as results of the spirantization of b and g. ʁ also only surfaces as a spirant version of ɢ, which is in itself an allophone of g that cannot be confused with existing sounds in Mongolian. Therefore the spirantization of the bilabial and velar phonemes, and of the uvular allophone of g, is not neutralizing.

Additional Comments

Note that the alveolar d does not pattern with the rest of the stop series and does not spirantize in intervocalic or any other contexts.

Notes on Difference from Kirchner's list

Lavoie (1996), cited in Kirchner (1998), lists two processes for Mongolian: the intervocalic spirantization of b g q to v ɣ x, and the loss of intervocalic ɣ. The first process is rewritten to more closely resemble the description in Lavoie's grammar source (Poppe 1970), adding parentheses around the uvular sounds because they are allophones of the velar g. The second process is omitted because it is not found in Poppe's grammar of Mongolian phonology.

3.85. Navajo

Language:	Navajo	*Dialect*:	Not specified
Phylum:	Na-Dene, Nuclear Na-Dene, Athapaskan-Eyak, Athapaskan, Apachean, Navajo-Apache, Western Apache-Navajo		
Inventory:	Lavoie	*Source*:	Kari 1976
Lenition(s):	1. x → h medially 2. ɣ j → Ø in verb prefixes		
Type:	1. Debuccalization 2. Loss		
Classification:	1. Not Neutralization 2. Not Neutralization		
Trends:	Two phonetic, but not phonological, neutralizations (x/h; and ɣ/j/Ø)		

Consonant Inventory

Table 94: Consonant Inventory of Navajo

									ʔ
	t^h	tl^h	ts^h	tʃ^h		k^h			
	t^ʔ	tl^ʔ	ts^ʔ	tʃ^ʔ		k^ʔ			
b	d	dl	dz	dʒ		g	g^j		
			s	ʃ		x			h
			z	ʒ		ɣ		ʁ	
m	n								
m^ʔ	n^ʔ								
	l̥								
	l								
w					j				
w^ʔ					j^ʔ				

Description

Process 1 is the weakening of x, a phoneme which is limited to word and syllable initial positions, to h when it is not word initial (3.75).

(3.75) x and h (p.10):

a.	[xai]	'winter'
b.	[nahasniih]	'I am trading it'

Process 2 is the deletion in verb prefixes of the voiced velar fricative (ɣ → Ø / V(#)_V) and the palatal glide (j → Ø / V#_i).

Discussion

The debuccalization of x to h leads to phonetic neutralization of the contrast between x and the existing h in the language. However, the existing h is limited to word-final position (3.76) while the h that alternates with x is limited to word initial and medial positions. Since the resulting and existing h's in the language do not appear in the same context, phonological neutralization between these two phonemes is avoided.

(3.76) existing h (p.10):

a.	[dah]	'up'
b.	[nahasniih]	'I am trading it'

The deletion of ɣ and j in verb prefixes interacts with another rule in the language: it occurs before the tone assimilation rule formalized in (3.77). Forms that illustrate the relationship between the deletion of ɣ and j and tone assimilation are given in (3.78). If deletion preceded tone assimilation it would bleed this rule, as is not the case.

(3.77) Tone assimilation ('[' refers to 'classifier + stem') (p.61):
V[-long] → V˦ / V˦ (#) C1_+C0[

(3.78) Ordering between ɣ, j deletion and tone assimilation (p.60):

a.	/na˦#ɣi+sh+l̥+te˧e˦l̥/	na˦a˦shte˦e˦l̥	Pgl 'carry an animate object'
b.	/ji˦+na˦#ji+l̥+kah/	ji˦ne˦i˦l̥kah	R3 'track it'

The loss of ɣ and j is confined to verb prefixes, which means neutralization of meaning distinctions is unlikely because the rule applies in such a limited environment that includes only function words. In addition to this, the fact that the consonants were deleted is evidenced in the spreading of the high tone (since the consonants were instrumental in providing the environment for this spreading before their deletion). Evidence of their presence, therefore, makes their absence upon deletion less likely to neutralize with Ø.

3.86. Nepali

Language: Nepali *Dialect*: Not specified

Phylum:	Indo-European, Indo-Iranian, Indo-Aryan, Northern zone, Eastern Pahari
Inventory:	Kirchner *Sources*: Bandu & Dahal 1971, Acharya 1991
Lenition(s):	1. p^h b^h k^h - ϕ β kx 2. b d → 'lax' b d / V_V 3. ɖ $ɖ^ɦ$ → ɽ $ɽ^ɦ$ non-initially 4. ts^h → h / V_V 5. h → Ø / V_V 6. b^h d^h g^h → b d g except word-initially (optional)
Type:	1. Spirantization 2. Laxing (not explained how this is manifested) 3. Flapping 4. Debuccalization 5. Loss 6. Deaspiration
Classification:	1. Not Neutralization 2. Not Neutralization 3. Not Neutralization 4. Not Neutralization 5. Limited Neutralization 6. Incomplete Neutralization
Trends:	Phonetic, but not phonological, neutralization between ts/h, and h/Ø.
Difference:	Process descriptions are corrected and regrouped

Consonant Inventory

Table 95: Phonemic Consonant Inventory of Nepali (Bandu & Dahal 1971)

p	t	ts	ʈ	k	
p^h	t^h	ts^h	$ʈ^h$	k^h	
b	d	dz	ɖ	g	
b^h	d^h	dz^h	$ɖ^h$	g^h	
		s			h
m	n			(ŋ)	
	ɾ				
	l				

The velar nasal ŋ is in parentheses because evidence for its phonemic status is weak.

Description

Process 1 is the spirantization of the aspirated stops p^{h} to ɸ non-initially, b^{h} to β word-finally, and k^{h} to kx intervocalically and word-finally. These are illustrated below.

(3.79) p^{h} → ɸ non-initially (Bandu & Dahal 1971: 19):

a.	/p^{h}ul/	[p^{h}ul]	'egg'
b.	/p^{h}iɾ/	[p^{h}iɾ]	'return'
c.	/səp^{h}a/	[səɸa]	'clean'
d.	/map^{h}/	[maɸ]	'excuse'
e.	/ap^{h}sos/	[aɸsos]	'sorry'

(3.80) b^{h} → β word-finally (Bandu & Dahal 1971: 19):

a.	/b^{h}iɾ/	[b^{ɦ}i^{ɦ}ɾ]	'cliff'
b.	/b^{h}at/	[b^{ɦ}a^{ɦ}t̥]	'cooked rice'
c.	/nib^{h}/	[niβ]	'be put out'
d.	/tsob^{h}/	[tsoβ]	'dip'

(3.81) k^{h} → kx intervocalically and word-finally (Bandu & Dahal 1971: 23):

a.	/k^{h}il/	[k^{h}il]	'sting'
b.	/k^{h}el/	[k^{h}el]	'play, game'
c.	/luk^{h}ure/	[lukxure]	'weakling'
d.	/lak^{h}/	[lakx]	'ten thousand'

Process 2 is the intervocalic laxing of b and d. There is no explanation as to how the laxing is phonetically manifested, but there is an illustration of where it occurs in the case of the bilabial (the laxed sound is indicated in bold).

(3.82) Regular vs. lax b (Bandu & Dahal 1971: 19):

a.	/bən/	[bən]	'forest'
b.	/bes/	[bes]	'fine, excellent'
c.	/seba/	[se**b**a]	'service'
d.	/kəbi/	[kə**b**i]	'poet'

Process 3 is the flapping of the retroflex voiced stops ɖ ɖ^{ɦ} to ɽ ɽ^{ɦ} non-initially. Bandu & Dahal make it clear that both the aspirated and non-aspirated members participate in the flapping, but only provide an example of the former (3.83).

(3.83) Allophones of ɖ^{ɦ} (Bandu & Dahal 1971: 22):

a.	/ɖ^{h}əl/	[^{ʔ}ɖ^{ɦ}ə^{ɦ}l]	'be fallen'

b. /gaḍʰa/ [gaɽʱaʱ] 'thick'

Process 4 is the intervocalic debuccalization of ts^h to h.[15] No examples are provided. Process 5 is the intervocalic loss of h in normal speech (3.84).

(3.84) Intervocalic loss of h (Bandu & Dahal 1971: 26):

a. /bəhiro/ [bəiro] 'deaf'

b. /məhə/ [məə] 'honey'

The final process listed for Nepali, process 6, is the optional deaspiration of b^h d^h g^h to b d g everywhere except word-initially.

Discussion

Process 1 is not neutralizing because ɸ β kx (or x) surface in the language only as outcomes of the spirantization in question and therefore cannot be confused with existing phonemes. Process 2 is also not neutralizing because, although it is not clear exactly how the laxing is manifested in b and d, the author specifies that the lax sounds are kept distinct from other sounds in the language. Process 3, again, results in sounds that only surface in the language as part of this process (ɽ ɽʱ) and therefore does not lead to neutralization with other phonemes.

Process 4 is also not neutralizing, but this time for a different reason: the debuccalization of ts to h phonetically neutralizes with the existing h in the language, but it occurs in the same environment where the existing h is lost (process 5) so phonological neutralization is avoided.

Process 5, the elision of intervocalic h, is described by Bandu & Dahal as normally resulting in a vowel cluster that wouldn't otherwise occur in the language. This suggests that the deleted h does not completely neutralize with Ø, because the "irregular" vowel cluster provides evidence of its presence. Hence, the chance of widespread contrast obliteration is minimized by the avoidance of complete neutralization between the presence and absence of the h. This process is therefore classified as limited neutralization.

Process 6 is a more complicated case because the non-initial deaspiration of b^h d^h g^h interacts with processes in the language listed here (i.e., word-final spirantization of b^h to β in process 1 and the laxing of intervocalic b d in process 2) as well as others, such as the nasalization of b d g in the vicinity of nasal vowels, which in that context should keep these stops distinct from the

[15] Acharya calls this sound a palatal stop (c) while Bandu & Dahal call it an alveolar affricate (ts). It isn't clear which one is right, but this question is not instrumental for our purposes as long as its labeling as one or the other is consistent.

deaspirated voiced stops. The extent of actual neutralization between the deaspirated b^h d^h g^h and the existing b d g is difficult to assess, and we know that contrast is maintained in various contexts (word-finally for b, intervocalically for b d, and around nasals for all three). However, it appears that some neutralization is inevitable, and for this reason, this process is classified as incomplete neutralization.

Notes on Difference from Kirchner's list

Kirchner (1998) lists the following 8 processes for Nepali: (1) p^h → ɸ / V_; (2) b → 'lax' b / V_V; (3) b^h → β / _#; (4) $ɖ^h$ → $ɽ^h$ except #_; (5) c^h → h / V_V; (6) k^h → x / V_; (7) h → Ø / V_C; and (8) b^h d^h g^h → b d g except word-initially. Almost every one of these processes is corrected to make it more consistent with the description in Kirchner's two language sources (Bandu & Dahal 1971 and Acharya 1991). In addition to this, processes 1, 3 and 6 are collapsed into one because they all affect similar sounds (aspirated stops) in the same manner (spirantization) and in relatively similar contexts.

3.87. Newari

Language:	Newari	*Dialect*:	Not specified
Phylum:	Sino-Tibetan, Tibeto-Burman, Himalayish, Mahakiranti, Newari		
Inventory:	Kirchner	*Source*:	Nanda 1971
Lenition(s):	h → Ø / V_V		
Type:	Loss		
Classification:	Not Neutralization		
Trends:	Phonetic, but not phonological, neutralization		

Consonant Inventory

Table 96: Consonant Inventory of Newari

p	t		tʃ	k	
p^h	t^h		tʃ^h	k^h	
b	d		dʒ	g	
b^h	d^h		dʒ^h	g^h	
		s			h
m	n			(ŋ)	
m^h	n^h				
	r				
	r^h				
	l				
	l^h				

The velar nasal in parentheses (ŋ) occurs in the Bhaktapur dialect. In Kathmandu ŋ occurs only as a variant release of long nasal vowels and as a variant of n before dorsal consonants.

Description

In Newari intervocalic h is either very lax or elided altogether.

Discussion

Nanda adds that "syllable initial h induces breathiness in a following vowel" (p.5). While there is phonetic neutralization between the elided h and Ø, the contrast between the existence and absence of h is maintained in the breathiness of the following vowel, and phonological neutralization is avoided.

3.88. Nez Perce

Language: Nez Perce *Dialect*: Not specified
Phylum: Penutian, Plateau Penutian, Sahaptin
Inventory: Lavoie *Source*: Aoki 1970
Lenition(s): 1. t k q - s x χ
2. h → Ø / C_
Type: 1. Spirantization
2. Loss
Classification: 1. Neutralization

	2. Limited Neutralization
Trends:	Phonetic, but limited phonological, neutralization (h/Ø)
Difference:	Process descriptions are corrected and regrouped

Consonant Inventory

Table 97: Phonemic Consonant Inventory of Nez Perce

p	t̪	t			k	q	ʔ
p^ʔ	t̪^{ʔ}	t^ʔ			k^ʔ	q^ʔ	
			s		x	χ	h
m	n̪						
m^ʔ	n̪^{ʔ}						
	l̪						
	l̪^{ʔ}						
	ɬ̪						
w				j			
w^ʔ				j^ʔ			

Description

Process 1 is the spirantization of the stops t k q to s x χ in the following environments: t is realized as s before n or w (3.85), k is realized as x before k, q, n, l or a phonological word juncture (3.86), and q is realized as χ before k, n, l or a phonological word juncture (3.87).

(3.85) t → s before n or w; t elsewhere (p.39):

a.	/'juʔt/	['juʔt]	'poor', 'pitiful'
b.	/'juʔtne/	['juʔsne]	'poor' (object case)
c.	/ʔuj'le:qt/	[ʔoj'la:qt]	'six'
d.	/ʔuj'le:qtwe/	[ʔoj'la:qswa]	'six' (of men)

(3.86) k - x (p.39):

a.	/temi'tekki/	[temi'texki]	'with huckleberry'
b.	/temit'knu:t/	[temit'xnu:t]	'without huckleberry'

(3.87) q - χ (p.39):

a.	/hitki'li:tuqkike/	[hitkili:toχkika]	'he went on back'
b.	/hi't^{ʔ}i:qne/	[hi't^{ʔ}i:χne]	'he spoke'

Process 2 is the loss of h when it appears after a consonant in normal speech.

(3.88) h → Ø / C_ (p.40):

a.	/hi'ne:shin/	[hi'ne:sin]	'tell them'
b.	/'hehen/	['hehen]	'stem', 'vine'

Discussion

Process 1 is neutralizing: the spirantized allophones of t k q are phonetically identical to existing phonemes in the language that can appear in the same context where spirantization takes place. Thus, s x χ that result from the spirantization of t k q neutralize with the existing s x χ in Nez Perce.

Process 2 constitutes a phonetic neutralization between the deleted h and Ø. However, the context where this process applies is actually quite limited because h can only be the first member of a CCC or CCCC cluster, and is only attested as the 2nd member of a CC cluster when the first member is one of 4 consonants (s, x, w and qˀ). Hence, the environment where h deletion applies is extremely limited, as is the potential for this process to neutralize contrast.

Notes on Difference from Kirchner's list

Lavoie (1996), cited in Kirchner (1998), lists 3 processes for Nez Perce: (1) c - s; (2) delete h post-C; and (3) k - x, q - ʁ. The first and third processes are rewritten to more closely resemble the way they are described in Lavoie's grammar source (Aoki 1970). They are also collapsed into one process because they involve the spirantization of voiceless stops in similar environments.

3.89. Nkore-Kiga

Language:	Nkore-Kiga	*Dialect*:	Not specified
Phylum:	Niger-Congo, Atlantic-Congo, Volta-Congo, Benue-Congo, Bantoid, Southern, Narrow Bantu, Central, J, Nyoro-Ganda (J.10)		
Inventory:	Lavoie	*Source*:	Taylor 1985
Lenition(s):	b → β / V_V		
Type:	Spirantization		
Classification:	Not Neutralization		

Consonant Inventory

Table 98: Consonant Inventory of Nkore-Kiga

p				(tʃ)		k	
b		d		(dʒ)		g	
	f		s	ʃ			h
	v		z	ʒ			
m					ɲ	ŋ	
		r					
w		(ɹ)			j		

Sounds in parentheses: tʃ and dʒ are allophones of k and g, and ɹ is an allophone of r.

Description

In Nkore-Kiga intervocalic b spirantizes to β.

Discussion

The bilabial spirant β only surfaces in the language as a result of the spirantization of b and therefore does not neutralize with existing sounds.

3.90. Northern Corsican

Language: Northern Corsican *Dialect*: Not specified
Phylum: Indo-European, Italic, Romance, Southern, Corsican
Inventory: Kirchner *Source*: Oftedal 1985
Add. Source: Dinnsen & Eckman 1977
Lenition(s):
1. p t k → b d g / V_V
2. b d g → β ð ɣ / V_V
3. f s → v z / V_V
4. v → w / V_V

Type:
1. Voicing
2. Spirantization
3. Voicing
4. Gliding

Classification:
1. Not Neutralization
2. Not Neutralization
3. Not Neutralization

	4. Not Neutralization
Trends:	Phonemic overlap between processes 1 & 2, and 3 & 4
Difference:	Process descriptions are corrected

Consonant Inventory

Not provided in either source

Description

The description in Oftedal of the intervocalic weakening in Corsican is very brief and incomplete. Dinnsen & Eckman, however, provide a more detailed discussion. The stops p t k are voiced in intervocalic context to b d g (process 1), and existing b d g in the language are spirantized in the same context to β ð ɣ (process 2).

(3.89) Voicing in Corsican (Dinnsen & Eckman 1977: 6):

a.	[peðe]	'foot'	[u beðe]	'the foot'
b.	[tengu]	'I have'	[u dengu]	'I have it'
c.	[kaza]	'house'	[a gaza]	'the house'

(3.90) Spirantization in Corsican (Dinnsen & Eckman 1977: 6):

a.	[bokka]	'mouth'	[a βokka]	'the mouth'
b.	[dente]	'tooth'	[u ðente]	'the tooth'
c.	[gola]	'throat'	[di ɣola]	'of throat'

The spirants f s are also voiced intervocalically to v z (process 3) and the existing v in the language is glided in the same context to w (process 4).

(3.91) Voicing of f s (Dinnsen & Eckman 1977: 6):

a.	[fele]	'faithful'	[u vele]	'the faithful one'
b.	[fyumi]	'river'	[u vyumi]	'the river'
c.	[sakku]	'bag'	[u zakku]	'the bag'
d.	[sole]	'sun'	[u zole]	'the sun'

(3.92) Gliding of v (Dinnsen & Eckman 1977: 7):

a.	[veni]	'come'
b.	[ajju ða wena]	'shall I come'
c.	[iddu weni]	'he comes'
d.	[vani a lliwornu]	'they go to Livorno'
e.	[du wa]	'where are you going?'

Discussion

All 4 processes discussed for Northern Corsican are allophonic; they do not neutralize contrasts in the language. What is interesting is that by themselves processes 1 and 3 would be considered neutralizing because they result in sounds that exist in the inventory of the language, but in the same context where these two voicing processes apply, so do spirantization and gliding, which affect the exact sounds that would have neutralized with the outcomes of voicing. That is, where intervocalic p t k are voiced to b d g, existing b d g are spirantized to β ð ɣ and neutralization between the outcomes of voicing and existing voiced stops is avoided. The same is true for the voiced f in the same environment as the gliding of v.

Notes on Difference from Kirchner's list

Kirchner (1998) lists the following as lenitions in Northern Corsican: "p t k - b d ɟ/g; b d ɟ/g - w Ø j w; ts tʃ - dz dʒ (contexts unclear)". Kirchner's source for Corsican grammar (Oftedal 1985) makes the context for the processes in question clear, but does not give a complete picture. Since Oftedal's primary grammar source is unavailable, a source unrelated to Kirchner's work is consulted on this language (Dinnsen & Eckman 1977).

3.91. Numic

Language:	Numic	*Dialect*:	Numic is a language family of 7 languages
Phylum:	Uto-Astecan, Northern Uto-Astecan, Numic		
Inventory:	Lavoie	*Source*:	Manaster Ramer 1993
Lenition(s):	*kk > k / V_V		
Type:	Degemination		
Classification:	Not Neutralization		
Trends:	Phonemic overlap (kk > k where k > h/x)		
Difference:	Process is corrected		

Consonant Inventory

No inventory of sounds is provided in the source. Numic is a subset of the Northern Uto-Astecan family.

Description

Geminates in Numic correspond to singletons in Tubatulabal and Luiseño: "In certain vocalic environments there is a regular correspondence between the Numic contrast -ɣ- (lenited from *-k-):-kk- and the Tubatulabal contrast -h-:-k-/-g-" (Manaster Ramer 1984, cited in Manaster Ramer 1993: 335).

Discussion

The author is concerned with the singleton vs. geminate contrast of Proto-Numic, and the preservation of this contrast in Tubatulabal and Luiseño. He shows that upon degemination of intervocalic kk, the contrast in the same context between k and kk is preserved: in Tubatulabal as h vs. k/g and in Luiseño as x vs. k/g. In other words, as kk shifted to k, in the same environment the existing k shifted to h/x (depending on the language), and the contrast between geminates and singletons was preserved. Hence, no neutralization resulted from this degemination.

Additional Comments

Although there is not much information in the source about the lenition of the singleton k, it is clear that the degemination of kk resulted in phonemic overlap because in the same environment where kk shifted to k, k shifted to something else (x, h, ɣ in the various languages).

Notes on Difference from Kirchner's list

Lavoie (1996), cited in Kirchner (1998), lists "k → x, h or ɣ / V_V" as a lenition process in Numic. Her source for Numic grammar (Manaster Ramer 1993) however, is not concerned with the shift of k to ɣ, h, or x, and in fact argues that this probably was not a result of any single process. Instead, the author is concerned with the preservation of the singleton vs. geminate contrast in Proto-Numic in Tubatulabal and in Luiseño. For this reason it is the degemination in question that is listed.

3.92. Nyawaygi

Language:	Nyawaygi	*Dialect*:	Not specified
Phylum:	Australian, Pama-Nyungan, Nyawaygic		
Inventory:	Kirchner	*Source*:	Dixon 1983

Lenition(s):	r → d / C_
Type:	Occlusivization
Classification:	Not Neutralization
Difference:	Sounds participating in process are corrected

Consonant Inventory

Table 99: Phonemic Consonant Inventory of Nyawaygi

b	(d)		ɟ	g
m	n		ɲ	ŋ
		ɽ		
	r			
	l			
w			j	

The author notes that the trill r and the alveolar stop d are allophones of the same phoneme.

Description

In reference to the r/d phoneme Dixon writes that it is "normally realized as an apico-alveolar stop [d] after a consonant and as an apico-alveolar trilled rhotic [r] in all other environments" (p.438).

Discussion

The author makes it clear that r and d belong to the same phoneme, so the alternation d - r does not lead to neutralization.

Additional Comments

The rhotic is the allophone of the r/d phoneme that surfaces in most environments, whereas d is limited to post consonantal contexts. For this reason the r is the one that traditionally would be considered the basic phoneme with d as its allophone, which implies the direction of the change is one of r to d. This is an example of occlusivization, which is normally considered fortition rather than lenition.

Notes on Difference from Kirchner's list

Kirchner (1998) lists the lenition process in Nyawaygi as flapping: d → ɾ except after a homorganic nasal. His language source, Dixon 1983 (Kirchner lists the publication year as 1979, but the grammar of Nyawaygi appears in Volume 3 of Dixon's work on Australian languages, which was published in 1983), specifies that d alternates with an "apico-alveolar trilled rhotic" and says nothing of a flap, or anything about a homorganic nasal.

3.93. Páez

Language: Páez *Dialect*: Not specified
Phylum: Paezan
Inventory: Lavoie *Source*: Gerdel 1985
Lenition(s): x - h
Type: Debuccalization
Classification: Not Neutralization
Difference: Process description is corrected

Consonant Inventory

Table 100: Consonant Inventory of Páez

p	t	tʲ	ts	tʃ			k		ʔ
b	d	dʲ					g		
ɸ			s	ʃ	ʃʲ		x	xʲ	
β			z	ʒ			ɣ		
m	n	nʲ							
	l	lʲ							
w						j			

Description

Gerdel describes the x - h alternation in a footnote:

> The voiceless velar fricative /x/ has fortis and lenis variants [x] and [h], [x] occurring between /k/ and stressed /i ɨ/ and in free fluctuation with [h] word initially, and [h] occurring elsewhere. (p.41)

Discussion

Although it is not clear whether Gerdel's inventory of sounds is limited to phonemes, the complete exclusion of h from this list makes it clear that the glottal fricative is not phonemic. Therefore, the alternation between the velar and glottal fricatives, since it results in a sound that does not otherwise surface in the language, does not neutralize contrasts with existing phonemes.

Notes on Difference from Kirchner's list

Lavoie (1996), cited in Kirchner (1998), lists the x - h alternation in Páez as occurring intervocalically. This is not the case according to Lavoie's grammar source (Gerdel 1985).

3.94. Panyjima (Mijaranypa)

Language:	Panyjima	*Dialect*:	Mijaranypa
Phylum:	Australian, Pama-Nyungan, South-West, Inland Ngayarda		
Inventory:	Kirchner	*Source*:	Dench 1991
Lenition(s):	1. ʈ → ɽ / V_V 2. r → ɾ or ɹ / V_V 3. no r / _#		
Type:	1. Flapping 2. Flapping/gliding 3. Distributional observation		
Classification:	1. Not Neutralization 2. Not Neutralization 3. Not Neutralization		
Difference:	Process descriptions are corrected		

Consonant Inventory

Table 101: Phonemic Consonant Inventory of Mijaranypa Panyjima

p	t̪	t	ʈ	ɟ	k
m	n̪	n	ɳ	ɲ	ŋ
		r			
	l̪	l	ɭ	ʎ	
w			ɻ	j	

Description

Process 1 is described by Dench as "the post-alveolar stop /ʈ/ typically occurs intervocalically as a retroflex flap" (p.131). Regarding the rhotic, Dench writes that "the alveolar rhotic /r/ is realized as a trill preceding a consonant. It does not occur in final position. In intervocalic position it is often realized as a single tap but perhaps more commonly occurs as an alveolar rhotic continuant" (p.132). Process 2 is, then, the intervocalic alternation between r and ɾ or ɹ, and process 3 is the fact that the rhotic does not occur in final position.

Discussion

Process 1 results in a retroflex flap (ɽ), which does not otherwise occur in Panyjima and therefore does not neutralize with existing sounds. Likewise, process 2 results in sounds (ɾ and ɹ) which do not otherwise occur in the language, and is also not neutralizing.

Process 3 is a distributional observation and does not involve an active alternation in the language. There is no reason to believe that the lack of word-final rhotics is a result of a rule in Panyjima phonology that prohibits this sound in said context. In any case, this phenomenon does not have the potential to neutralize contrast because it does not involve an alternation.

Notes on Difference from Kirchner's list

Kirchner (1998) lists the first process as ʈ - ɾ, which would suggest neutralization between intervocalic ʈ and r since both would resulted in ɾ. Kirchner's grammar source for Panyjima (Dench 1991) however, makes it clear that the retroflex alternates with a retroflex, and not an alveolar, flap (ɽ not ɾ). Kirchner also adds a note to the distributional observation that r cannot appear word-finally, asking in parentheses whether it neutralizes to ʈ. However, not only does Dench not describe anything in Panyjima grammar that would support this view, but the restriction of a sound to certain contexts does not in itself have the potential to lead to neutralization.

3.95. Pawnee (South Band & Skiri)

Language:	Pawnee	*Dialect*:	South Band & Skiri
Phylum:	Caddoan, Northern, Pawnee-Kitsai, Pawnee		
Inventory:	Lavoie	*Source*:	Parks 1976
Lenition(s):	1. h → Ø / #_ (in South Band dialect, with some exceptions)		

	2. ɾ → Ø / _#
Type:	1. Loss
	2. Loss
Classification:	1. Neutralization
	2. Neutralization
Difference:	The description of the 2nd process is corrected

Consonant Inventory

Table 102: Phonemic Consonant Inventory of South Band & Skiri Pawnee

p	t	ts	k	
		s		h
	ɾ			
w				

Description

Process 1 is that in the South Band dialect of Pawnee word-initial h is lost (3.93), although there are a few lexical exceptions to this rule (3.94). Additional rules in Pawnee delete h in several, more complex, environments; these are not discussed here.

(3.93) Word-initial loss of h in South Band (p.38):

a.	/ha:t+u/	a:tu	'tongue'
	cf. /ti+ha:t+ta:hi:s+uk+Ø/		tihacta:his 'he's thirsty' (lit. dry-tongued)
b.	/ha:ka+u/	a:kau	'mouth'
c.	/haturа:ɾ+u/	atura:ɾu	'road'

(3.94) Exceptions to h-loss (p.38):

a.	hawa	'also'
b.	he	'and'

Process 2 is the loss of a stem-final ɾ when it occurs in word-final position. He adds:

> This rule applies to a large number of stems, especially descriptive verbs. nearly half of the descriptive stems that have been recorded have a base final ɾ which occurs phonetically when a suffix, e.g., the subordinate form, occurs following it; but when the verb is in its dependent form, the stem is not followed by any overt marker (i.e., is followed by Ø) and base final ɾ is deleted. (p.49)

(3.95) ɾ deletion in descriptive verbs (a, b) and elsewhere (c, d, e) (p.49):

a.	/ti+ta:ka:ɾ+Ø/	tita:ka:ʔa	'it is white'
	cf., /ɾa+ku+ta:ka:ɾ+uk+Ø/	rakuta:ka:ɾu	'to be white'
	/ta:ka:ɾ/		'to be white'
b.	/ti+takipi:ɾ+Ø/	titakipi:ʔi	'it is closed'
	cf., /takipi:ɾ/		'to be closed'
c.	/ti+ku+uɾ+kiɾiɾ+Ø/	tikuhkiɾi	'I shivered'
	cf., /uɾ… kiɾiɾ/		'to shiver, shake'
d.	/ta+t+ha+i:ɾ+Ø/	tataʔi	'I put it in a liquid'
	cf., /ha+i:ɾ/		'to put in a liquid'
e.	/tsawi:iɾ/	tsawi:ʔi	'Chawi'
	cf. /tsawi:iɾ+ɾu/	tsawi:ɾiɾu	'among the Chawi'

Discussion

Both deletion processes listed for Pawnee appear to be neutralizing, obliterating the distinction between the deleted sounds (h and ɾ) and Ø.

Notes on Difference from Kirchner's list

Lavoie (1996), cited in Kirchner (1998), lists the two processes for Pawnee as "delete h word-initially" and "delete r word-finally". I follow her grammar source (Parks 1976) more closely in listing the loss of h as affecting only one dialect, and the word-final loss of ɾ rather than r.

3.96. Pipil (Cuisnahuat)

Language:	Pipil	*Dialect*:	Cuisnahuat
Phylum:	Uto-Aztecan, Southern Uto-Aztecan, Aztecan, General Aztec, Pipil		
Inventory:	Lavoie	*Source*:	Campbell 1985
Lenition(s):	w → h / _#, _C (in one dialect)		
Type:	Debuccalization		
Classification:	Neutralization		

Consonant Inventory

Table 103: Consonant Inventory of Cuisnahuat Pipil

p		t	ts	tʃ		k	
						k^w	
(b)		(d)				(g)	
	(f)		s	ʃ			h
m		n					
		(ɾ)					
		(r)					
		l					
w					j		

Sounds in parentheses may not be phonemic; b d g occur only in words of Spanish origin and alternate between stops and fricatives depending on context.

Description

In the Cuisnahuat dialect of Pipil a syllable-final w becomes h in word-final or pre-consonantal position.

(3.96) Examples of w - h (p.34):

a.	kuwa	'to buy'	-kuh-ki	'bought'
b.	puwa	'to count'	-puh-ki	'counted'

Discussion

Although the Proto-Nahua *h is lost in some environments, it is retained in word-final and pre-consonantal positions (and one other context, at the end of a reduplicated first syllable, which isn't significant to us). Therefore, the h that alternates with w can appear in the same position as an existing h in the language, which may lead to neutralization.

3.97. Proto-Ainu

Language:	Proto-Ainu	*Dialect*: Not specified
Phylum:	Language Isolate	
Inventory:	Lavoie	*Source*: Vovin 1993
Lenition(s):	*g > h	
Type:	Debuccalization	

Classification:	Not Neutralization
Difference:	Process descriptions are corrected and regrouped

Consonant Inventory

Table 104: Reconstructed Consonant Inventory of Proto-Ainu

p	t			k	(q)	
	d			g		
		s				h
						(ɦ)
m	n					
	ɾ					
			j			

Description

Vovin reconstructs Proto-Ainu *g from the h in dialects of Ainu, which suggests that at some point in the language's history there was a shift from g to h.

Discussion

Although diachronic data makes functional determinations rather difficult, the author argues in this particular case that the contrast between Proto-Ainu *h and *g (>h) was preserved. Vovin claims that the existing h and the h that resulted from *g did not occur in the same context, which indeed would avoid neutralization.

Notes on Difference from Kirchner's list

Lavoie (1996), cited in Kirchner (1998), lists two processes for Proto-Ainu: g > h and delete h, both as probably V_V. Lavoie's language source (she cites Vavin, but it is Vovin 1993) makes absolutely no reference to the two processes as occurring intervocalically, which is why the context is omitted. The second process is described by Vovin as *h > h/Ø, but very few details are provided and it is not possible to determine whether this process led to the neutralization of any contrasts. Therefore, as I do with other diachronic processes for which contrast details are scarce, I omit it from my corpus.

3.98. Proto-Bantu

Language: Proto-Bantu *Dialect*: Not specified
Phylum: Niger-Congo, Atlantic-Congo, Volta-Congo, Benue-Congo, Bantoid, Southern
Inventory: Kirchner *Source*: Halle & Clements 1983
Add. Source: Greenberg 1948 (grammar source for Halle & Clements)
Lenition(s): b d g → β l ɣ except following a homorganic nasal
Type: Spirantization
Classification: Not Neutralization

Consonant Inventory

Table 105: Consonant Inventory of Proto-Bantu

p	t			k
b	d			g
		s		
(β)		z		(ɣ)
m	n			ŋ
	(l)			
(w)			(j)	

The inventory is compiled based on data in both sources listed above. In parentheses: β l ɣ are specifically listed as in complementary distribution with b d g (Halle & Clements 1983) and w j may also not be phonemic (Greenberg 1948).

Description

In Proto-Bantu b/β d/l and g/ɣ are variants of each other where the first member of each pair (b d g) occurs before nasals and the second (β l ɣ) elsewhere. As an illustration of this process compare the transcription in Greenberg, who does not mark the spirants but instead concentrates on tones, and in Halle & Clements where the phonetic, rather than phonemic, transcription is used (3.97).

(3.97) Compare Greenberg 1948 and Halle & Clements 1983:

Greenberg	Halle & Clements	
ba˩de˥ (p.202)	βale (p.55)	'two'
ki˧gɛ˩ (p.205)	kiɣa (p.55)	'eyebrow'
ta˧be˩ (p.207)	taβe (p.55)	'twig'

Discussion

There are few details provided in both sources about the process in question, but one thing is made clear: b d g and β l ɣ belong to the same phonemes and are in complementary distribution with each other. Since β l ɣ exist in the language only as members of the b d g phonemes, they do not obliterate contrast with other phonemes.

3.99. Proto-Germanic

Language:	Proto-Germanic	*Dialect*:	Not specified
Phylum:	Indo-European, Germanic		
Inventory:	Kirchner	*Sources*:	Meillet 1970, Prokosch 1933
Lenition(s):	1. Stage 1: b^h d^h g^h > β ð ɣ 2. Later: p^h t^h k^h > f θ x		
Type:	1. Spirantization 2. Spirantization		
Classification:	1. Not Neutralization 2. Not Neutralization		

Consonant Inventory

Not provided in the source

Description

The spirantization of the voiced aspirated stops to β ð ɣ, and the voiceless aspirated stops to f θ x, are both part of Grimm's law which describes the sound shift from Indo European to Proto-Germanic.

Discussion

While neither of Kirchner's sources provides details, the case of Grimm's law is well known as an example of a sound shift where the contrast within the inventory of sounds was preserved. That is, the voiced and voiceless spirant series entered the language only as part of this shift and therefore did not neutralize with existing phonemes.

3.100. Ptolemaic Greek

Language: Ptolemaic Greek *Dialect*: Not specified
Phylum: Indo-European, Greek
Inventory: Lavoie *Source*: Teodorsson 1977
Lenition(s): g > ɣ
Type: Spirantization
Classification: Not Neutralization

Consonant Inventory

Table 106: Phonemic Consonant Inventory of Ptolemaic Greek

p	t		k	
pʰ	tʰ		kʰ	
b	d		g	
		s		h
		(z)		
m	n		(ŋ)	
	r			
	l			

Sounds in parentheses: z is an allophone of s which occurs before tonal consonants, and ŋ is an allophone of n which occurs before a velar stop.

Description

Teodorsson provides no more details except that g shifted to ɣ.

Discussion

According to the inventory of phonemes, ɣ only surfaced in the language as part of the g phoneme and therefore could not have obliterated any contrasts.

3.101. Punjabi

Language: Punjabi *Dialect*: Not specified
Phylum: Indo-European, Indo-Iranian, Indo-Aryan, Central zone, Panjabi
Inventory: Kirchner *Source*: Gill & Gleason Jr. 1969
Lenition(s): 1. Geminates occur only after centralized vowels

	2. $p^h \rightarrow ϕ$ / V_V, _#
Type:	1. Distributional property
	2. Spirantization
Classification:	1. Not Neutralization
	2. Not Neutralization
Difference:	Description of process 1 is corrected; one process is omitted

Consonant Inventory

Table 107: Consonant Inventory of Punjabi

p		t		ʈ	c	k	
p^h		t^h		$ʈ^h$	c^h	k^h	
b		d		ɖ	ɟ	g	
	(f)		s	ʂ			h
			(z)				
m		n		ɳ	ɲ	ŋ	
				ɽ			
		r					
		l		ɭ			
w					j		

The two sounds in parentheses occur only in loan words from Persian or Arabic.

Description

Regarding gemination in Punjabi, Gill & Gleason Jr. write that "all consonants occur geminated except /r ɭ ɳ ɽ ɲ ŋ h j w/. Geminates are prolonged and more fortis. [...] Geminates occur only after centralized vowel." (p.14-15).

Process 2 is the alternation p^h - ϕ: "/p^h/ is commonly heard as a fricative [ϕ], particularly between vowels and in word-final position" (p.10). This is said to be more common in faster and less formal speech.

Discussion

The first lenition listed for Punjabi is not an active phonological process. The restriction of geminates to certain contexts, in this case after centralized vowels, does not in itself imply that there is degemination everywhere else in the language. A distributional property of the language does not have the potential to result in neutralization because it does not involve any alternations that may affect contrast.

The alternation p^h - ɸ also does not result in neutralization because both sounds exist in Punjabi only as part of this phoneme (represented by p^h in the inventory). Therefore it does not have the potential to interfere with other phonemes and obliterate any contrasts.

Notes on Difference from Kirchner's list

Kirchner (1998) lists 3 processes for Punjabi: (1) all stops are lenis in medial position after centralized vowels; (2) after peripheral vowels, geminates degeminate; and (3) p^h → ɸ in casual speech, particularly /V_V and /_#. I consult the same grammar source (Gill & Gleason Jr. 1969) which mentions nothing of medial lenition, and in fact specifies that after centralized vowels is the only place where geminates, not usually considered 'lenis', can occur. For this reason I omit the first of Kirchner's processes. I also correct the second process to reflect more closely the fact that the authors never discuss degemination in any vowel contexts, but only mention the distributional property of geminates.

3.102. Purki

Language:	Purki[16]	*Dialect*: Not specified
Phylum:	Sino-Tibetan, Tibeto-Burman, Himalayish, Tibeto-Kanauri, Tibetic, Tibetan, Western	
Inventory:	Kirchner	*Source*: Rangan 1979
Lenition(s):	1. d̪ → ð / V_V	
	2. d → ɾ / V_V	
	3. No allophones of -ant sounds, even velars, but note no /ɢ/	
Type:	1. Spirantization	
	2. Flapping	
	3. Distributional property, of sorts	
Classification:	1. Not Neutralization	
	2. Not Neutralization	
	3. Not Neutralization	
Trends:	Flapping of alveolar possibly patterns with spirantization	
Difference:	Description of process 2 is corrected	

[16] Although their inventories are very similar, I assume that Purki the language is not the same as the Purki dialect of Balti (§3.12) based on work by Rangan 1975. Balti is spoken in Pakistan, whereas Purki is a language of India. Both stem from the same Phylum.

Consonant Inventory

Table 108: Phonemic Consonant Inventory of Purki

p		t̪	t	ts	tʃ		k	q	
p^h		t̪^{h}	t^h	ts^h	tʃ^h		k^h		
b		d̪	d		dʒ		g		
	(f)			s	ʃ			χ	h
				z	ʒ			ʁ	
m			n			ɲ	ŋ		
			r						
			l						
w						j			

The sound in parentheses, f, is a 'marginal' phoneme.

Description

Process 1 is the fact that the dental voiced stop d̪ alternates with the dental fricative ð in intervocalic position (3.98). Process 2 is the alternation between d and ɾ in intervocalic position (3.99).

(3.98) Variations of d̪ (p.17):

a.	/a:za:d̪i/	[a:za:ði]	'freedom'
b.	/d̪ika/	[d̪ika]	'here'

(3.99) Variations of d (p.17-18):

a.	/kabadi/	[dʌbʌɾi]	'kabadi'
b.	/dim/	[dim]	'trunk (of a tree)'

Listed as process 3 is Kirchner's observation that there is an asymmetry in Purki's inventory and phonology. First, he observes that no other voiced stops except for the dental and alveolar alternate with a flap or spirant intervocalically, and second that the uvular voiced stop is not part of the inventory.

Discussion

Process 1 is not neutralizing because both alternating sounds, d̪ and ð, occur in Purki only as part of this phoneme (represented by d̪ in the inventory). Therefore there is no potential for contrast obliteration between this and other phonemes in the language.

Similarly, process 2 is not neutralizing because d and ɽ both only occur in the language as part of the d phoneme, and therefore cannot affect contrasts with other phonemes.

The observation about the asymmetry in the inventory of Purki's voiced stops and their behaviors does not have the potential to neutralize any contrasts in the language because it does not include any alternations (in fact, it emphasizes the lack of alternation).

Notes on Difference from Kirchner's list

Kirchner lists as process 2 for Purki the intervocalic alternation between d and r. Had this been true, this alternation would have been neutralizing because r is phonemic. Kirchner's grammar source (Rangan 1979) however, describes the alternation as involving the alveolar voiced stop d and a retroflex flap ɽ, which is not phonemic in Purki.

3.103. Quechua

Language:	Quechua	*Dialect*:	Not specified
Phylum:	Quechuan		
Inventory:	Kirchner	*Source*:	Whitley 1978
Add. Source:	Bills et al. 1969 (Whitley's grammar source)		
Lenition(s):	k q → x χ in coda		
Type:	Spirantization		
Classification:	Not Neutralization		

Consonant Inventory

Table 109: Phonemic Consonant Inventory of Quechua (Bills et al. 1969)

p	t		tʃ		k	q	
p^h	t^h		tʃ^h		k^h	q^h	
p^ʔ	t^ʔ		tʃ^ʔ		k^ʔ	q^ʔ	
		s					h
m	n			ɲ			
	r						
	l			ʎ			
w				j			

Description

Bills et al. describe the spirantization in Quechua as "phonemic /k/ and /q/ are realized phonetically as fricatives, velar [x] and post-velar or uvular [χ] respectively, when they occur in syllable-final position" (p.xix).

(3.100) Spirantization of k and q (Bills et al. 1969: xix):

a.	/ʎikʎa/	[ʎixʎa]	'small shawl'
b.	/tʃiqnin/	[tʃεχniŋ]	'he hates'

Discussion

The spirant versions of k and q only surface in the language as results of the spirantization process in question and therefore do not neutralize with existing sounds.

3.104. Russian

Language:	Russian	*Dialect*:	Not specified
Phylum:	Indo-European, Slavic, East		
Inventory:	Kirchner	*Source*:	Wade 1992
Lenition(s):	1. Voicing assimilation to following obstruent		
	2. Optional degemination word-internally		
Type:	1. Voicing		
	2. Degemination		
Classification:	1. Neutralization		
	2. Not Neutralization		

Consonant Inventory

Table 110: Consonant Inventory of Russian

p	p^j			t	t^j	ts		tʃ		k	k^j
b	b^j			d	d^j					g	g^j
		f	f^j			s	s^j	ʃ	ç	x	x^j
		v	v^j			z	z^j	ʒ			
m	m^j			n	n^j						
				r	r^j						
				l	l^j						
									j		

Description

Process 1 is that obstruents in Russian assimilate in voicing to the sound they precede (3.101, 3.102). An exception is v, which does not induce voicing in the preceding sound (3.103).

(3.101) Voicing assimilation: devoicing (p.12):

a.	gubka	['gupkə]	'sponge'
b.	zags	[zaks]	'registry office'
c.	vxodit	['fxodjɪt]	'goes in'

(3.102) Voicing assimilation: voicing (p.13):

a.	futbol	[fu'dbol]	'football'
b.	k domu	['gdomu]	'towards the house'

(3.103) Exception (p.13):

tvoj	[tvoj]	'your'

Process 2 refers to Wade's report that double consonants within a stem are optionally pronounced as single.

(3.104) Medial double consonants (p.14):

a.	grammatika	[m]	'grammar'
b.	gruppa	[pp or p]	'group'

Discussion

The assimilation in voicing in Russian neutralizes the contrast between voiced and voiceless sounds.

Process 2 is most likely not an active process in Russian, but instead an explanation of how to read the language. That is, gemination in Russian is not part of the phonology but part of the orthography, in which case speakers do not encode geminates in the lexicon and do not actively degeminate these phonemes in certain environments. In native Russian words gemination is not contrastive so regardless of whether there is active degemination word medially or not, degemination would not lead to neutralization because no contrast in the language depends on this feature.

3.105. Saek

Language: Saek *Dialect*: Not specified

Phylum: Tai-Kadai, Kam-Tai, Be-Tai, Tai-Sek, Sek
Inventory: Kirchner *Source*: Hudak 1993
Lenition(s): ɣ ~ g / #_
Type: Occlusivization
Classification: Not Neutralization
Difference: Process description is corrected

Consonant Inventory

Table 111: Phonemic Consonant Inventory of Saek

p		t		c	k	ʔ
p^h		t^h		c^h	k^h	
b		d			(g)	
	f		s			h
					ɣ	
m		n		ɲ	ŋ	
		r				
		l				
w	ʋ			j		

Hudak places the voiced velar g in parentheses to indicate that it is an allophone of ɣ.

Description

In Saek the voiced velar spirant ɣ is slowly being replaced by the stop g: "the initial voiced velar stop g is usually pronounced by younger speakers, with older speakers favoring the voiced spirant ɣ" (p.xxvi). In (3.105) the tone is mid-level with a slight fall and a glottal constriction.

(3.105) ɣ ~ g (p.xxxi):

Old Generation	Young Generation	
ɣaw^{tone}	gaw^{tone}	'rice'

Discussion

Hudak makes it clear that the two voiced velars, g and ɣ, are members of the same phoneme. Therefore the alternation between them does not interfere with other phonemes in the language and does not lead to neutralization.

Additional Comments

The directionality of the alternation between the velar stop and spirant is clear: the younger generation prefers the stop to the spirant, which suggests the spirant is indeed slowly being replaced by a stop. This is an example of occlusivization rather than spirantization, and as such is not an example of lenition.

Notes on Difference from Kirchner's list

Two minor changes are made: first, Kirchner (1998) cites the language source for Saek as "Gedney 1993", but this grammar is authored by Hudak and is based on Gedney's work. Second, the process is listed as ɣ ~ g / #_ (rather than g ~ ɣ) in accordance with the author's description of the process as occlusivization which occurs word-initially.

3.106. Sanuma

Language:	Sanuma	*Dialect*:	Not specified
Phylum:	Yanomam		
Inventory:	Lavoie	*Source*:	Borgman 1990
Lenition(s):	1. p t ts k → b d dz g / V_V (optional)		
	2. d dz → ɾ z / V_V in fast speech (optional)		
	3. h → Ø / V_V (optional)		
Type:	1. Voicing		
	2. Spirantization		
	3. Loss		
Classification:	1. Not Neutralization		
	2. Not Neutralization		
	3. Neutralization		
Trends:	Flapping patterns with spirantization (process 2)		
Difference:	Process descriptions are corrected and regrouped		

Consonant Inventory

Table 112: Phonemic Consonant Inventory of Sanuma

p	t	ts	k	
	tʰ			
		s		h
m	n			
	l			
w				

Description

The voiceless stops and affricate p t ts k alternate with their voiced counterparts b d dz g intervocalically (process 1).

(3.106) Intervocalic voicing in Sanuma (p.220):

a.	ipa	[ipa] or [iba]	'my'
b.	hute	[hute] or [hude]	'heavy'
c.	hatsa	[hatsa] or [hadza]	'deer'
d.	ãka	[ãka] or [ãga]	'tongue'

In addition to the intervocalic voicing, the alveolars t ts are also optionally realized as ɾ z in fast speech (process 2). Since this occurs in the same environment where t ts are voiced the shift is formalized as d dz - ɾ z.

(3.107) Spirantization of t (p.220):

kutiata [kutiaɾa] or [kuɾiaɾa] 'canoe'

Process 3 is that word-medial h is often lenis or sometimes lost completely.

(3.108) Lenition of h (p.221):

ohi [ohi] or [oi] 'hungry'

Discussion

Process 1 is not neutralizing because the voiced versions of the stops p t ts k surface in the language only as alternants to the voiceless sounds and are not members of any other phonemes. Similarly, process 2 is not neutralizing because ɾ and z only surface in Sanuma as members of the alveolar stop phonemes and do not interfere with other phonemes.

Process 3, the optional lenition and sometimes loss of intervocalic h, appears to be neutralizing. Although it is not clear how much contrast is actually obliterated, especially since the loss is the extreme version of an optional alternation, the fact remains that there is neutralization between h and Ø.

Notes on Difference from Kirchner's list

Lavoie (1996), cited in Kirchner (1998), lists the first process for Sanuma as p t ts k → b ɾ dz g / V_V. I divide the voicing of stops, and subsequent spirantization of the alveolars (adding the affricate) into two separate processes.

3.107. Sawai

Language:	Sawai	*Dialect*:	Not specified
Phylum:	Austronesian, Malayo-Polynesian, Central-Eastern, Eastern Malayo-Polynesian, South Halmahera-West New Guinea, South Halmahera, Southeast		
Inventory:	Lavoie	*Source*:	Whisler 1992
Lenition(s):	d - ɾ		
Type:	Flapping		
Classification:	Not Neutralization		
Difference:	Process description and citation information are corrected		

Consonant Inventory

Table 113: Phonemic Consonant Inventory of Sawai

p		t			k
b		d			g
	f		s		
m		n			ŋ
		l			
w				j	

Description

Whisler reports that d and ɾ alternate: the flap occurs intervocalically in roots, as the first of two consonants in clusters, syllable finally, occasionally word-initially, and sometimes in unpredictable environments (3.109a-c), whereas d occurs elsewhere (3.109d-g). The rest of the stop series does not participate in similar alternations.

(3.109) Distribution of d and ɾ (p.9-10):

a.	[ɾakɛn]	'close'
b.	[n-dɛɾɛɾɛ]	'3s-to follow'
c.	[m-diɾkɛ]	'2s-to move'
d.	[deko]	'friend'
e.	[mdi]	'tree trunk'
f.	[dodego]	'sugar press'
g.	[t-dɛ]	'1p-to go along'

Discussion

Since the flap is not listed as phonemic in Sawai it is safe to assume that it surfaces in the language only as part of the d phoneme, and therefore, the d - ɾ alternation does not neutralize contrasts with other phonemes.

Notes on Difference from Kirchner's list

Lavoie (1996), cited in Kirchner (1998), lists the d - ɾ alternation in Sawai as occurring intervocalically and post-vocalically. I add a more detailed environment for this process according to the description in Lavoie's grammar source (Whisler 1992).

3.108. Sekani

Language:	Sekani *Dialect*: Not specified
Phylum:	Na-Dene, Nuclear Na-Dene, Athapaskan-Eyak, Athapaskan, Canadian, Beaver-Sekani
Inventory:	Lavoie *Source*: Hargus 1985
Lenition(s):	s l̥ ç x ʍ → z l ʝ ɣ w when noun/stem-initial is prefixed
Type:	Voicing
Classification:	Not Neutralization
Trends:	Phonetic, but not phonological, neutralization
Difference:	Process description and citation are corrected

Consonant Inventory

Table 114: Phonemic Consonant Inventory of Sekani

p	t	tl̥	ts	tʃ		k	ʔ
pʰ	tʰ	tl̥ʰ	tsʰ	tʃʰ		kʰ	
	t'	tl̥'	ts'	tʃ'		k'	
						kʰʷ	
			s		ç	x	h
			z		ʝ	ɣ	
m	n						
	l̥						
	l						

Description

Regarding continuant voicing in Sekani, Hargus writes:

> In general, there is no contrast between voiced and voiceless stem-initial fricatives in underlying representations. Surface voicing is predictable. Following Rice (1983a, b) I assume that stem-initial fricatives are underlyingly voiceless in this position, with a voicing rule providing surface voicing in appropriate environments. (p.269)

Hargus reports that voiceless initials of noun and postposition stems (s l̥ ç x ʍ) are voiced (z l ʝ ɣ w) when prefixed or preceded by a nominal possessor or object.

(3.110) Examples of continuant voicing in Sekani (p.270-271):

a.	xa˧s	'planning tool'	səɣa˧se˧	'my planning tool'
b.	çən	'song'	səʝəne˧	'my song'
c.	xa˧z	'windfall roots'	tse ɣa˧z-e	'Old Friend Mt. (*roots* stem)'

Discussion

The voiced sounds participating in the s l̥ ç x ʍ → z l ʝ ɣ w alternation are included in the inventory of phonemes which suggests that continuants in Sekani are indeed contrastive with respect to voicing. Based on this, the alternation in question does result in phonetic neutralization between the voiceless continuants and their voiced counterparts. However, Hargus specifically notes that in the position where the alternation takes place the fricatives are not contrastive with respect to voicing. That is, while in other contexts in the language a voicing distinction between continuants indicates a distinction in phonemes, in the context where continuant voicing is described (stem-initially upon prefixation) voicing is not distinctive because only the voiceless phonemes can occur. So all voiced continuants in this context are representations of the voiceless phonemes, and there is never a question as to whether a continuant represents a voiced phoneme or a voiced version of a voiceless one. Hence phonological neutralization is avoided.

Notes on Difference from Kirchner's list

Lavoie (1996), cited in Kirchner (1998), suggests that the alternation in question takes place in all intervocalic contexts in the language. This is not the case, and I adhere to Lavoie's grammar source (Hargus 1985) in describing the context as limited to prefixed stem-initials.

3.109. Senoufo

Language:	Senoufo	*Dialect*:	Not specified
Phylum:	Niger-Congo, Atlantic-Congo, Volta-Congo, North, Gur, Senufo		
Inventory:	Lavoie	*Source*:	Mills 1984
Lenition(s):	1. p t k → b d g in unstressed medial position 2. b d g → β ɾ ɣ in unstressed medial syllables		
Type:	1. Voicing 2. Spirantization		
Classification:	1. Not Neutralization 2. Not Neutralization		
Trends:	Phonemic overlap between processes 1 & 2; flapping patterns with spirantization		

Consonant Inventory

Table 115: Phonemic Consonant Inventory of Senoufo

p		t			k	kp
b		d			g	gb
	f		s			
	v		z			
m		(n)		(ɲ)	(ŋ)	
		l				
w				j		

The nasals in parentheses are "marginal phonemes" (p.92).

Description

In Senoufo p t k are voiced to b d g in unstressed medial position (process 1), and in the same context b d g are spirantized to β ɾ ɣ (process 2). It isn't clear whether kp and gb participate in these alternations.

Discussion

The spirantization of b d g in the same context where p t k are voiced serves to avoid neutralization between the voiced and voiceless phonemes in the language. Thus, neither of the processes is neutralizing because all medial (in unstressed syllables) voiced stops represent the voiceless phonemes, and spirants represent only the voiced stop phonemes. This is an example of phonemic overlap between the voiceless and voiced series.

3.110. Serbo-Croatian

Language:	Serbo-Croatian	*Dialect*:	Not specified
Phylum:	Indo-European, Slavic, South, Western		
Inventory:	Kirchner	*Source*:	Partridge 1988
Lenition(s):	Regressive voicing assimilation in obstruent clusters		
Type:	Voicing assimilation		
Classification:	Neutralization		

Consonant Inventory

Table 116: Consonant Inventory of Serbo-Croatian

p		t	t^j	ts	tʃ		k	
b		d	d^j		dʒ		g	
	f			s	ʃ			h
	v			z	ʒ			
m		n	n^j					
		r						
		l	l^j					
						j		

Description

Partridge describes the voicing assimilation in Serbo-Croatian as

> When any two or more of the fourteen 'paired' consonants occur in juxtaposition in a consonant cluster, mutation takes place and all become voiced or unvoiced according to the last. (p.20)

(3.111) Examples of voicing assimilation (p.20):

a.	rob	'slave'	vs.	ropstvo	'slavery'
b.	sladak	'sweet'	vs.	slatka	'sweet (fem. sg.)'

Discussion

As in other Slavic languages where this type of voicing assimilation is common, there is neutralization between the voiced and voiceless phonemes.

3.111. Shapsug (Düzce)

Language:	Shapsug	*Dialect*:	Düzce
Phylum:	North Caucasian, Northwest, Circassian		
Inventory:	Kirchner	*Source*:	Smeets 1984
Lenition(s):	Articulatory force varies in clusters		
Type:	Unrelease		
Classification:	Not Neutralization		
Trends:	Phonetic, but not phonological, neutralization		

Consonant Inventory

Table 117: Phonemic Consonant Inventory of Düzce Shapsug

p	t̪	t		tʃ	c	k	q		ʔ
$p^{ʔ}$	$t̪^{ʔ}$	$t^{ʔ}$			$c^{ʔ}$	$k^{ʔ}$			
						k^{w}	q^{w}		$ʔ^{w}$
$p^{wʔ}$	$t̪^{wʔ}$					$k^{wʔ}$			
b	d̪	d		(dʒ)	ɟ	g			
						g^{w}			
ɸ			s	ʃ	ç	x	χ	ħ	h
			$s^{ʔ}$	$ʃ^{ʔ}$					
$ɸ^{w}$				$ʃ^{w}$	$ç^{ɣ}$		$χ^{w}$		
				$ʃ^{wʔ}$					
			z	ʒ	ʝ	ɣ	ʁ		
				$ʒ^{w}$			$ʁ^{w}$		
m		n							
		r							
		ɬ		ɮ					
		$ɬ^{ʔ}$							
w					j				

Description

Smeets writes that

> In clusters, a voiceless final member has relatively weak aspiration, whereas a non-final member is pronounced with less articulatory force than a single consonant. The non-final member usually lacks aspiration or glottal release of its own. (p.115)

Discussion

Both voicing and aspiration appear to play a role in the voiced/voiceless opposition for laryngeals in the language. Word-finally aspiration is reported as more distinctive than voicing, so a weakening in the aspiration of a voiceless final member in a cluster does potentially threaten contrast. However, while weaker, the aspiration is still reported as audible so the contrast is not obliterated.

The non-final member of a cluster, on the other hand, may lack aspiration and is partially devoiced. This only occurs across morpheme boundary, but "consonant sequences which display an internal morpheme boundary are made up of two or three obstruents, the first of which itself constitutes an allomorph of a personal prefix" (p.116). Thus, the phonetic neutralization that the non-final member of a cluster undergoes does not lead to phonological neutralization because it is limited to one function word in the language.

3.112. Shina

Language:	Shina	*Dialect*:	Not specified
Phylum:	Indo-European, Indo-Iranian, Indo-Aryan, Northwestern zone, Dardic, Shina		
Inventory:	Kirchner	*Source*:	Rajapurohit 1983
Lenition(s):	1. b d g → ƀ đ ǥ / V_V, r_V 2. ɖ → ɽ / V_V 3. r → ɾ / (b, d, g)_, _V[short]		
Type:	1. Spirantization 2. Spirantization/Flapping 3. Flapping		
Classification:	1. Not Neutralization 2. Not Neutralization 3. Not Neutralization		
Trends:	Flapping somewhat patterns with spirantization		
Difference:	Process descriptions are corrected and regrouped		

Consonant Inventory

Table 118: Phonemic Consonant Inventory of Shina

p		t	ts	tʃ	ʈ	ʈʂ		k	
pʰ		tʰ	tsʰ	tʃʰ	ʈʰ	ʈʂʰ		kʰ	
b		d		dʒ	ɖ			g	
			s	ʃ	ʂ			x	h
	v		z	ʒ	ʐ			ɣ	
m		n			ɳ			ŋ	
		r							
		l							
							j		

Description

Process 1 is the slight spirantization of b d g (to ƀ đ ǥ) when they occur between two vowels or between r and a vowel. Process 2 is the intervocalic alternation between ɖ and ɽ. And process 3 involves the flapping of r to ɾ after b d g and also before a short vowel.

Discussion

Process 1 involves the alternation between the b d g and a slightly fricativized version of these sounds (ƀ đ ǥ). This is significant because although the bilabial and alveolar spirants β and ð are not phonemic in the language, the velar voiced spirant ɣ is. However, the author makes a point of distinguishing between the slightly fricativized velar ǥ and the spirant ɣ, emphasizing that the two do not neutralize phonetically. Thus, none of the sounds involved in this alternation are members of phonemes other than b d g, and therefore, the alternation does not obliterate any contrasts.

Process 2 is quite similar to the spirantization of b d g, especially in view of other languages in the present study where flapping patterns with spirantization. This process is listed separately because its context is not identical to that of the spirantization of b d g. The retroflex flap ɽ is not phonemic in Shina, and therefore, its alternation with ɖ does not interfere with other phonemes in the language: it is not neutralizing.

Process 3, the flapping of r to ɾ, is also not neutralizing because the alveolar flap is not phonemic and surfaces in the language only as part of the r phoneme.

Notes on Difference from Kirchner's list

Kirchner (1998) lists the lenitions in Shina as the spirantization of b d g and the flapping of retroflex d. I correct this list to more closely resemble the description in Kirchner's language source (Rajapurohit 1983). Mainly, I specify that ɖ flaps to a retroflex ɽ, and the alveolar flap ɾ alternates with the trill.

3.113. Shoshone (Tümpisa)

Language:	Shoshone	*Dialect*:	Tümpisa
Phylum:	Uto-Aztecan, Northern Uto-Aztecan, Numic, Central		
Inventory:	Kirchner	*Source*:	Dayley 1989a
Add. Source:	Shimkin 1949, Dayley 1989b		
Lenition(s):	1. p t c k k^w → ɸ θ ç x x^w / V_ [voiceless]V		
	2. p t c k k^w → β ð ʝ ɣ $ɣ^w$ / V_ [voiced]V		
	3. t → ɾ / [back]V_V		
	4. p t c k k^w → b d ɟ g g^w / N_V		
	5. m n → w̃ j̃ / V_V		
	6. ʔ h → Ø / V_V		

Type: 1. Spirantization
2. Voicing
3. Flapping
4. Voicing
5. Gliding
6. Loss

Classification: 1. Not Neutralization
2. Not Neutralization
3. Not Neutralization
4. Not Neutralization
5. Not Neutralization
6. Not Neutralization

Trends: Flapping patterns with spirantization (processes 3); phonetic, but not phonological, neutralization (glottals/Ø in process 6)

Difference: All process descriptions are corrected and regrouped

Consonant Inventory

Table 119: Phonemic Consonant Inventory of Tümpisa Shoshone (Dayley 1989a)

p	t		c	k	(ʔ)
				kʷ	
		s			h
m	n			ŋ	
				ŋʷ	
w			j		

The glottal ʔ is in parentheses because there is good reason to believe it is not phonemic. It is not clear whether by c Dayley meant the palatal stop or the alveolar affricate ts, however, contrast matters discussed below are not affected by either decision.

Description

The stops p t c k kʷ are spirantized in intervocalic context, to ɸ θ ç x xʷ before voiceless vowels (process 1) and to β ð ʝ ɣ ɣʷ before voiced vowels (process 2), although at times the voiced and voiceless spirants are in free variation. The alveolar t only spirantizes to θ/ð after front vowels, and after back vowels it flaps to ɾ/ɾ̥ (process 3).

(3.112) Some examples of spirantization of stops (Dayley 1989a: xxviii-xxix):

a. wisipin [wiʃiɸi̥] 'thread'

b.	tapettsi	[taβettʃi̥]	'sun'
c.	citoohin	[ciðo:hi̥]	'push'
d.	peti^{-gem}	[peði] ~ [peθi̥]	'daughter'[17]
e.	kati^{-gem}	[kaɾi] ~ [kaɾ̥i̥]	'sit'
f.	mi?akʷa!	[mĩ?aɣʷa] ~ [mĩaɣʷa]	'go away!'

When in a voiced environment other than intervocalic (e.g., between a nasal and a vowel) the stops p t c k kʷ are voiced to b d ɟ g gʷ (process 4). As in the case of the post-spirantization voicing (process 2) the voiced and voiceless counterparts are often in free variation regardless of their environment.

(3.113) Voicing of stops (Dayley 1989a: xxvii):

a.	isapuŋku	[iʃaβũŋgu] ~ [iʃaβũŋku]	'dog'
b.	timpin	[tĩmbi] ~ [tĩmpi]	'rock'
c.	itiintin	[iɾĩindi] ~ [iɾĩinti]	'heat, hot place'

Process 5 is the intervocalic gliding of the nasals m and n to the nasalized voiced and voiceless glides w̃/w̥̃ and j̃/j̥̃. In the case of the alveolar n the alternation does not take place after back vowels.

(3.114) Gliding of nasals (Dayley 1989a: xxviii):

a.	nimi	[niw̃i] ~ [niw̥̃i]	'person, people'
b.	wihnu	[w̃ĩhj̃ũ] ~ [w̃ĩhj̃u̥]	'then'[18]

Process 6 is the free variation between the presence and absence of the glottals ? h in intervocalic position.

(3.115) ? h - Ø (Dayley 1989a: xxix):

a.	mi?akʷa!	[mĩ?aɣʷa] ~ [mĩaɣʷa]	'go away!'
b.	so?oppitin	[sɔ?ɔp:iɾi] ~ [sɔ:p:iɾi]	'much, many'

[17] A word that is marked as $^{-gem}$ means it ends with a geminating segment, which is one of the three possible "final features" in the language (the others being nasalizing and preaspirating). What Dayley (1989a) refers to as "final features" are the unique characteristics of the only consonant-like sounds that can be found word-finally in Tümpisa Shoshone.

[18] Assuming this is not a typo in the source, it appears that an intervening h does not block the intervocalic gliding of the nasals. This is interesting, but not instrumental to my study.

Discussion

Since neither spirantization nor voicing are contrastive in the language, processes 1 and 2 are not neutralizing: the sounds that participate in the spirantizing alternations, whether they are voiced or voiceless spirants, are not part of any other phonemes in Tümpisa Shoshone. Process 3, the flapping of the alveolar sounds after back vowels, is also not neutralizing because flaps are not phonemic and therefore only represent the alveolar phoneme. Similarly, process 4 is not neutralizing because, as already mentioned, voicing is not contrastive and therefore the voiced stops can only represent the voiceless stop phonemes and do not obliterate distinction with other phonemes in the language.

Process 5 is not neutralizing because, in alternating with glides, m n retain their nasal quality, maintaining the contrast between existing glides (w j) and the ones that alternate with the nasals (w̃ j̃).

Process 6 involves the optional loss of the glottals in intervocalic position. The loss of h does not lead to any neutralization because "a distinct syllabic pulse remains between the two vowels" (Dayley 1989a: xxiv). Thus, there is an audible sound that distinguishes the presence of the h phoneme from Ø, and neutralization is avoided.

The loss of ʔ also avoids neutralization because, despite the fact that Dayley classifies this sound as a phoneme, it does not appear to be one. Shimkin (1949) specifically lists this sound as non-phonemic in Shoshone, although it is possible that the Tümpisa dialect may have a different system of contrasts. However, the data in Dayley (1989a, b) do not seem to support his analysis of ʔ as phonemic.

The ʔ can only appear intervocalically, which is where it is optionally dropped. However, intervocalic ʔ is also subject to insertion rules: Dayley (1989b) reports that a phrase-final marker of a glottal stop + voiceless vowel are often added word-finally to words spoken in isolation and at the end of phrases (3.116). This suggests that the alternation between ʔ and Ø in intervocalic context may also be one of insertion rather than deletion, and there is nothing in Dayley (1989a, b) that contradicts such an analysis. Epenthetic matter, which ʔ may very well be, is not normally awarded phonemic status.

(3.116) ʔ + V insertion (Dayley 1989b: 418-419):

a.	papi	paβiʔi̥	'older brother'
b.	tɨmmu	ɾ̃ɨmmũʔu̥	'enemy'
c.	ma	mãʔḁ	'with (instrument)'
d.	haa	haaʔḁ	'yes'

More importantly, a thorough examination of Dayley's 416 page Tümpisa Shoshone-English dictionary produced only two examples of minimal pairs (3.117).[19]

(3.117) Minimal pairs (Dayley 1989a):

a.	toi	'go to the toilet (from Engl. *toilet*)'	vs.
	toʔi	'pipe' (p.289)	
b.	moʔo	'hand, fist'	vs.
	-moo-	'decimal (sɨːmootin '10', wahamootin '20', etc.)' (p.105)	

Note that neither of these is truly a minimal pair: their distinction does not fully depend on ʔ because toi in (3.117a) is a foreign word and moo in (3.117b) is never used in isolation.

I conclude that, although it is possible that minimal pairs that show the contrast between ʔ/Ø in Tümpisa do exist, Dayley's classification of ʔ as phonemic, based on his own data, is not justified. Furthermore, if ʔ truly is phonemic, its functional load must be so small that there is no potential for widespread neutralization resulting from its deletion. With this in mind, process 5 does not obliterate meaning distinctions either in the case of the h or the ʔ.

Notes on Difference from Kirchner's list

Kirchner (1998) lists 4 processes for Tümpisa Shoshone: (1) P T K - β ð ɣ; (2) ɾ/ð conditioned by backness of preceding V; (3) m n ŋ^w → w̃ j̃/ɾ̃ w̃ except utterance initially and in geminates; and (4) ʔ h → Ø (morphologized context, phonological factors unclear). All of these processes are rewritten to conform to the description in Kirchner's grammar source (Dayley 1989a). I divide Kirchner's first process that includes spirantization and voicing of stops into 2: the spirantization of stops (process 1), the spirantization as well as voicing of stops (process 2), and the voicing of stops (process 4). Kirchner's 2nd process is rewritten as the flapping of t (process 3). Kirchner's 3rd process is corrected to the gliding of m and n. Dayley says nothing of the gliding of ŋ^w, which had Kirchner been correct in his description, would have neutralized with the gliding

[19] Since ʔ can only occur V_V it was easy to look through the dictionary for minimal pairs because there was no possibility of the minimal pairs not following each other in the alphabetic list (i.e., since no words start with ʔ, there was no need to compare words listed under different letters). In fact, Dayley (1989a: xxv) specifies that "Glottal stop ʔ is ignored in alphabetical order unless two forms are identical, except that one contains glottal stop and the other does not. In this case the form with glottal stop follows the form without it (e.g., toi would come before toʔi)".

of m. Kirchner's final process had only one correction: the context for glottal lost is actually made quite clear in the source as intervocalic.

3.114. Somali

Language:	Somali	*Dialect*:	Not specified
Phylum:	Afro-Asiatic, Cushitic, East, Somali		
Inventory:	Lavoie	*Source*:	Armstrong 1964
Lenition(s):	b d ɖ g → β ð ɽ ɣ / V_V, especially after stress		
Type:	Spirantization		
Classification:	Not Neutralization		
Trends:	Flapping patterns with spirantization		

Consonant Inventory

Table 120: Consonant Inventory of Somali

		t					k	q		ʔ
b		d		dʒ	ɖ		g			
	f		s	ʃ			x		ħ	h
									ʕ	
m		n								
		r								
		l								
w						j				

Description

The voiced stops b d ɖ g often spirantize to β ð ɽ ɣ in intervocalic environment, especially after a stressed syllable.

(3.118) Spirantization of the stops:[20]

a.	ˈlaba	ˈlaβa	'two'
b.	ˈɖaban	ˈɖaβan	'cheek'
c.	ˈbadag	ˈbaðag	'goose'
d.	magaaˈlada	magaaˈlaða	'the city'
e.	ˈtiɖi	ˈtiɽi	'she said'
f.	ˈɖaɖab	ˈɖaɽab	'dream'
g.	ˈsagaal	ˈsaɣaal	'nine'

[20] In the 1964 edition of Armstrong's work there are no page numbers.

h. ˈnin ˈraga ˈnin ˈraɣa ‘brave man’

Discussion

Neither the spirants β ð ɣ nor the retroflex flap ɽ are included in the inventory of consonants for Somali which means they are not phonemic and only surface as part of the alternation with the voiced stops. Therefore, there is no neutralization between phonemes.

3.115. Sotho (Southern)

Language: Sotho *Dialect*: Southern
Phylum: Niger-Congo, Atlantic-Congo, Volta-Congo, Benue-Congo, Bantoid, Southern, Narrow Bantu, Central, S, Sotho-Tswana (S.30), Sotho, Southern
Inventory: Kirchner *Source*: Doke & Mofokeng 1957
Lenition(s): 1. l → d / _[+high]V
2. b > β
Type: 1. Delateralization
2. Spirantization
Classification: 1. Not Neutralization
2. Not Neutralization
Difference: 1st process corrected, 2nd process rewritten according to available source

Consonant Inventory

Table 121: Consonant Inventory of Southern Sotho

pʰ	pʃʰ		tʰ	tɬʰ	tsʰ	tʃʰ		kʰ	kxʰ	
p’	pʃ’		t’	tɬ’	ts’	tʃ’		k’		
b	bʒ		d			dʒ				
	fʃ	f			s	ʃ		(x)		h
		(v)				ʒ				
m			n				ɲ	ŋ		
m̩			n̩				ɲ̩	ŋ̩		
			r							
			l							
			l̩							
			ɬ							
w							j			

The nasals and lateral with the diacritic below are syllabic (e.g., n̩, m̩). The labial-alveolar affricates (pʃʰ, pʃ', bʒ, fʃ) are called "compound sounds". The sounds in parentheses are found only in foreign acquisitions.

Description

Process 1 is that in Southern Sotho, before the high vowels i or u, l alternates with d.

(3.119) l → d before high vowels (p.19):

a.	bala	'read'
b.	-badile	'read (perf. stem)'

Process 2 is the spirantization of b to β. No context is listed for this process, and it appears that the source is describing a diachronic change b > β, and that all instances of the bilabial voiced stop phoneme are realized as the fricative. The authors add that the "Southern Sotho b is the phonological equivalent of ɓ (implosive) in the Nguni group" (p.14).

Discussion

Process 1 is not neutralizing because the only one of the alternating sounds (l and d) which can occur before the high vowels is the d, and according to the authors, the d in this context is always the one that participates in the l - d alternation. Therefore, whether d's that do not participate in the alternation exist in the language or not, they cannot be confused with the phoneme that does alternate with l because all d's in this environment represent the l/d phoneme.

Process 2 is also not neutralizing because β only surfaces in the language as a representation of the b phoneme and therefore does not obliterate contrast with other phonemes.

Additional Comments

Doke & Mofokeng specifically call the first process the delateralization of l (to d) before high vowels, and not the lateralization of d to l. Therefore, it is probably not a case of lenition.

Notes on Difference from Kirchner's list

Kirchner (1998) lists the first process in Southern Sotho as d → l / _[-high]V. Kirchner's source (Doke & Mofokeng 1957), however, specifies that d replaces l before high vowels (l → d / _[+high]V). The second process in Kirchner's inventory is a general spirantization in certain vowel-conditioned environments. This process is cited from Grammont (1939), a source which was not available to me, and is therefore omitted from my inventory. Instead, I include a slightly less general version of the same lenition process (i.e., spirantization) which is discussed in the one source for Southern Sotho available to me (b - β).

3.116. Southern Tati (Chali)

Language:	Southern Tati	*Dialect*: Chali
Phylum:	Indo-European, Indo-Iranian, Iranian, Western, Northwestern, Talysh	
Inventory:	Kirchner	*Source*: Yar-Shater 1969
Lenition(s):	1. q → ʁ non-initially 2. h → Ø non-initially (in loanwords)	
Type:	1. Spirantization and voicing 2. Loss	
Classification:	1. Not Neutralization 2. Not Neutralization	
Trends:	Phonetic, but not phonological, neutralization (h/Ø)	
Difference:	Description of 1st process is corrected; processes are regrouped	

Consonant Inventory

Table 122: Consonant Inventory of Chali Southern Tati

				tʃ			q	(ʔ)
p^h		t^h				k^h		
b		d		dʒ		g		
	f		s	ʃ			χ	h
							($χ^w$)	
	v		z	(ʒ)				
m		n						
		ɾ						
		l						
					j			

Parentheses in the consonant inventory are not explained in the source.

Description

Process 1 involves the articulation of q, which ranges from plosive to fricative ʁ. In general, in contexts other than initial (3.120a) the ʁ variant is used.

(3.120) Variants of q (p.34):

a.	qeleft	[qeˈleft]	'throat'
b.	lɑqa	[laˈʁæ]	'branch'
c.	uʃqɑlden	[uʃʁalˈden]	'to press'
d.	etɑq	[eˈtaʁ]	'room'

Process 2 is the loss of h in pre-consonantal non-initial position. This mostly affects loanwords from Persian, but may also affect native words.

Discussion

Process 1 is not neutralizing because the ʁ only surfaces in Chali as part of the q phoneme and therefore does not have the potential to interfere with other phonemes and to obliterate contrast.

Process 2 is also not neutralizing because the loss of h is accompanied by compensatory lengthening of the preceding syllable. This means that, although the loss of h phonetically neutralizes the contrast between the h phoneme and Ø, phonological contrast is maintained by shifting to the length of the preceding syllable. Thus, the lengthened syllable makes the existence of the h phoneme (which, after loss is phonetically Ø) apparent.

Notes on Difference from Kirchner's list

Kirchner (1998) includes the devoicing of obstruents in his list of Chali processes but his grammar source (Yar-Shater 1969) never mentions any such process. Therefore this process is omitted from my corpus. The sounds participating in the alternation listed as process 1 are also corrected to more closely resemble the description in the Yar-Shater.

3.117. Southern Tati (Eshtehardi)

Language:	Southern Tati	*Dialect*: Eshtehardi
Phylum:	Indo-European, Indo-Iranian, Iranian, Western, Northwestern, Talysh	
Inventory:	Kirchner	*Source*: Yar-Shater 1969

Lenition(s):	v → w / (ɵ, o, ɒ)_
Type:	Gliding
Classification:	Not Neutralization
Difference:	Split analysis for this dialect and Xiarah

Sound Inventory

Table 123: Consonant Inventory of Eshtehardi Southern Tati

p		t		tʃ		k	q	(ʔ)
b		d		dʒ		g		
	f		s	ʃ			χ	h
	v		z	(ʒ)				
m		n						
		ɾ						
		l						
					j			

Table 124: Vowel Inventory of Eshtehardi Southern Tati

i			u
	e	ɵ	o
		a	ɒ

Description

Yar-Shater writes that "the labio-dental v has a bilabial allophone [w] which occurs, in the available material, only after ɵ, o, and ɒ" (p.43).

(3.121) Gliding in Eshtehardi (p.43):

a.	gɵw	'ox'
b.	qowwat	'strength'
c.	ɒwa	'water'

Discussion

The gliding of v in Eshtehardi is not neutralizing because the glide w only surfaces in the language as part of the alternation with v, and does not interfere with any other phonemes.

Notes on Difference from Kirchner's list

Kirchner (1998) lists the gliding of v in the Eshtehardi and Xiarah dialects of Southern Tati as one process in his inventory. However, these two dialects have different consonant inventories, and bearing in mind that other dialects of Southern Tati are listed as separate languages, I separate these two as well (see §3.119 for Xiaraji).

3.118. Southern Tati (Takestani)

Language: Southern Tati *Dialect*: Takestani
Phylum: Indo-European, Indo-Iranian, Iranian, Western, Northwestern, Talysh
Inventory: Kirchner *Source*: Yar-Shater 1969
Lenition(s): g → ɣ / o_o
Type: Spirantization
Classification: Not Neutralization

Consonant Inventory

Table 125: Consonant Inventory of Takestani Southern Tati

p		t		tʃ		k	q	(ʔ)
b		d		dʒ		g		
	f		s	ʃ			χ	h
	v		z	(ʒ)				
m		n						
		ɾ						
		l						
					j			

Description

Yar-Shater writes "g has an especially retracted articulation when preceded and followed by o. Two of my informants regularly had [ɣ] in this position" (p.40).

(3.122) Spirantization of g (p.40):
a. [moˈɣo] 'must'
b. [oɣoˈrem] 'that I lift'
c. [moˈɣostem] 'I wanted'
d. [oɣorˈdæn] 'to lift'

Discussion

The spirantized articulation of the velar does not lead to obliteration of any contrasts because the ɣ only surfaces in the language as part of its alternation with g, and therefore, does not affect other phonemes.

3.119. Southern Tati (Xiaraji)

Language:	Southern Tati	*Dialect*:	Xiaraji
Phylum:	Indo-European, Indo-Iranian, Iranian, Western, Northwestern, Talysh		
Inventory:	Kirchner	*Source*:	Yar-Shater 1969
Lenition(s):	v → w / (ɵ, o, a)_		
Type:	Gliding		
Classification:	Not Neutralization		
Difference:	Split analysis for this dialect and Eshtehardi		

Consonant Inventory

Table 126: Consonant Inventory of Xiaraji Southern Tati

p		t		tʃ		k	q	(ʔ)
b		d		dʒ		g		
	f		s	ʃ			χ	h
							χ^{w}	
	v		z	(ʒ)				
m		n						
		ɾ						
		l						
(w)					j			

Description

Yar-Shater describes w in Xiaraji as an allophone of v which does not occur initially, and is attested only after a, o, and ɵ.

(3.123) Distribution of w in Xiaraji (p.45):

a. awjɒɾ 'water-distributor'
b. migow 'must'
c. kɵwa 'blue, blue grey'

Discussion

The glide w is reported to not be contrastive, and to belong only to the v phoneme. Therefore, the gliding of v does not obliterate any contrasts in the language.

Notes on Difference from Kirchner's list

The Southern Tati dialects Eshtehardi and Xiarah, listed together in Kirchner's (1998) inventory, are separated into two sections (see §3.117 for Eshtehardi).

3.120. Spanish (Andalusian)

Language: Spanish *Dialect*: Andalusian
Phylum: Indo-European, Italic, Romance, Italo-Western, Western, Gallo-Iberian, Ibero-Romance, West Iberian, Castilian
Inventory: Kirchner *Source*: Romero-Gallego 1995
Lenition(s): b d g → β̞ ð̞ ɣ̞ except #_, l_, or N_ (N = homorganic nasal)
Type: Spirantization
Classification: Not Neutralization

Consonant Inventory

Table 127: Consonant Inventory of Castilian Spanish (Harris & Vincent 1987)

p		t		tʃ		k
b		d				g
	f	(θ)	s			x
m		n			ɲ	
		ɾ				
		r				
		l			(ʎ)	
w					j	

No inventory of Andalusian Spanish consonants was available in the source so the inventory for Castilian Spanish is provided here. The sounds in parentheses are specific to the Castilian dialect of Spanish. It is not clear what, if any, differences may exist between the standard and the Andalusian dialects of Spanish.

Description

In this dialect of Spanish there is non-initial spirantization of voiced stops (b d g) except after a homorganic nasal or a lateral. The reduction is greater following a lower vowel and in faster speech.

Discussion

The spirantization of voiced stops is discussed at length by Romero-Gallego specifically with regards to whether it is allophonic or not. In particular, this dialect shows a rare preservation of the distinction between voiced approximants (β̞ ð̞ ɣ̞) which result from rule (3.124), and voiced fricatives (β ð ɣ) which result from the interaction of rules (3.124) and (3.125) in the contexts sb sd sg.

> (3.124) Spirantization (p.2):
> /b d g/ (except initial or after homorganic nasal or /l/) → [β̞ ð̞ ɣ̞]
>
> (3.125) s-aspiration (p.2):
> Reduction/loss of /s/ in syllable coda

Romero-Gallego emphasizes that, although approximants of the β̞ ð̞ ɣ̞ kind are rare in the languages of the world, in Andalusian they are not only present, but are also contrastive with the voiced fricatives β ð ɣ. Thus, the especially rare situation arises where two complete sets of continuants exist in the phonemic inventory of this dialect.

Romero-Gallego's study shows that Andalusian fricatives and approximants are indeed distinct. Romero-Gallego also reports that duration, rather than constriction degree, provides the main cues for the distinction between the two sets of continuants. The vital information for our purpose is that the spirantization of b d g does not neutralize any contrasts in the language, and in fact, preserves a rare contrast between approximants and fricatives.

3.121. Spanish (Mexico City)

Language:	Spanish	*Dialect*:	Mexico City
Phylum:	Indo-European, Italic, Romance, Italo-Western, Western, Gallo-Iberian, Ibero-Romance, West Iberian, Castilian		
Inventory:	Kirchner	*Source*:	Harris 1969
Lenition(s):	b d g → β ð ɣ in most environments, with exceptions		
Type:	Spirantization		
Classification:	Not Neutralization		

Trends: This process is part of phonemic overlap (p → b, b → β); one process omitted

Consonant Inventory

Not provided in the source, but in the previous section (§3.120: Andalusian Spanish) I provide an inventory of Castilian Spanish consonants.

Description

The voiced stops b d g are spirantized everywhere except word or utterance initially and after a homorganic nasal or lateral.

Discussion

Despite the fact that Harris does not include much information on contrasts in the language, enough is known about Spanish to conclude that this process is not neutralizing because β ð ɣ only surface as part of this process, and therefore, cannot be confused with existing sounds.

Additional Comments

Although Kirchner (1998) does not list it, the source also discusses the voicing of stops in a similar environment where voiced stops are spirantized. Hence, the spirantization of b d g helps avoid neutralization with the outcome of the voicing of p t k. The voicing of p t k to b d g, and spirantization of b d g to β ð ɣ, constitute phonemic overlap.

Notes on Difference from Kirchner's list

Kirchner (1998) includes a second process for Mexico City Spanish: r → ɾ except word-initially. Much has been written about the history of Spanish and the trill-tap contrast in this language, which is the current manifestation of the historical geminate-singleton contrast. Most significantly, it is known that there was a diachronic strengthening of initial trills in Ibero-Romance languages rather than a non-initial weakening (Penny 2002). Hence, despite the fact that Harris (1969) reports this as a process in the language, the non-initial weakening of trills in Mexico City Spanish is omitted from my corpus.

3.122. Tahltan (Iskut)

Language:	Tahltan	*Dialect*: Iskut
Phylum:	Na-Dene, Nuclear Na-Dene, Athapaskan-Eyak, Athapaskan, Tahltan-Kaska	
Inventory:	Lavoie	*Source*: Nater 1989
Add. Source:	Hardwick 1984 (one of Nater's grammar sources)	
Lenition(s):	Stem-initial and stem-final consonants are voiced	
Type:	Voicing	
Classification:	Not Neutralization	
Difference:	Process description is corrected	

Consonant Inventory

Table 128: Consonant Inventory of Iskut Tahltan (Nater 1989)

									ʔ	
	t̪ʰ	tʰ	tl̥ʰ		ʈʰ	cʰ	kʰ	qᵡ		
	t̪ʔ	tʔ	tl̥ʔ		ʈʔ	cʔ	kʔ	qʔ		
							kʰʷ			
							kʷʔ			
b	d̪	d	dl		ɖ	ɟ	g	ɢ		
							gʷ			
		θ		s		ç	x	χ	h	hʲ
							xʷ			
		ð		z		ʝ	ɣ	ʁ		
m		n̥								
		n								
		nʔ								
		l̥								
		l								
w						j				

The placement of qᵡ in the chart above is questionable.[21]

[21] The author writes χ as a superscript indicating it should be grouped with other aspirated stops. However, Nater also uses superscripts for affricates (e.g., tˡ), so it is possible that qχ should be listed as an affricate instead of as a uvular version of the aspirated stops.

Description

Upon affixation the stem-initial and stem-final consonants are voiced.

(3.126) Examples of stop voicing in Tahltan (Nater 1989: 31-32):

a.	xən̥	'(a) song'	ˈt^{s}ejˈɣəne	'funeral dirge'
			ʔesˈɣəne	'my song'
b.	baːt̪	'gloves, sleeves'	ˈʔeːc^{h}ˈbaːd̪e	'shirt-sleeves'
			ʔesˈbaːd̪e	'my gloves'
c.	xət̪^{h}	'(a) house'	ʔesˈɣəd̪e	'my house',
d.	l̥et̪	'smoke'	ʔesˈled̪e	'my smoke'
e.	keˈk^{ʔ}əʈ^{h}	'box'	ʔesgeˈk^{ʔ}əɖe	'my box'
f.	seːk	'saliva'	ʔesˈzeːge	'my saliva'
g.	l̥iʔ	'dog'	ʔesˈliʔe	'my dog'

Discussion

The voicing alternations in Tahltan are not neutralizing because voicing is not contrastive in the language. Nater does not provide contrast information in his inventory of sounds (that is, he includes both phonemes and surface representations) but he does call the voiced alternants "allophones" of the voiceless ones, suggesting the voiced and voiceless versions of each sound belong to the same phonemes. Additional support for the contrastive role voicing plays in the language comes from Hardwick 1984 as well as personal communication with John Alderete and Sharon Hargus, both researchers of Athapaskan languages.

Notes on Difference from Kirchner's list

Lavoie (1996), cited in Kirchner (1998), lists the voicing in Tahltan as an intervocalic and word-final alternation between a set of 7 sounds. I adhere more closely to Lavoie's grammar source (Nater 1989) which makes it clear that the alternation is a much more general one which affects all consonants, and specifies a different context for this phenomenon.

3.123. Taiwanese

Language:	Taiwanese	*Dialect*:	Not specified
Phylum:	Sino-Tibetan, Chinese		
Inventory:	Kirchner	*Source*:	Hsu 1995

Add. Sources:	Lin 1989, Zhang 1983 (in Hsu 1995)
Lenition(s):	p t k → β ɾ ɣ / V_V
Type:	Spirantization and voicing
Classification:	Not Neutralization
Trends:	Flapping patterns with spirantization
Difference:	Process description is corrected

Consonant Inventory

Table 129: Phonemic Consonant Inventory of Taiwanese (Hsu 1995)

p	t	ts	k	ʔ
pʰ	tʰ	tsʰ	kʰ	
b		dz	g	
		s		h
m	n		ŋ	
	l			

Description

Word-final p t k, when they find themselves in intervocalic position upon affixation, are spirantized to β ɾ ɣ. The degree of spirantization is variable, but it is always accompanied by voicing.

(3.127) p t k - β ɾ ɣ (p.5):[22]

a.	apa	>	aβa	'little box'
b.	kʰuta	>	kʰuɾa	'little hole'
c.	tika	>	tiɣa	'bamboo'

Discussion

The spirants β ɾ ɣ only surface in Taiwanese as outcomes of the alternation discussed here and don't affect other phonemes in the language. Therefore, no neutralization results from the spirantization and voicing of p t k.

Notes on Difference from Kirchner's list

Kirchner (1998) lists the alternation in Taiwanese as P T K → β ð ɣ except word initially. This description is corrected to adhere more closely to Kirchner's grammar source (Hsu 1995). Hsu uses capital P T K to indicate non-initial

[22] Hsu does not indicate tones in her data because they are irrelevant to her study.

sounds, but as this process does not occur word-initially there is no need to use this formalism. Also, Hsu reports that the alveolar alternates with ɾ and not ð, and that the process takes place intervocalically and not just non-initially.

3.124. Tamazight Berber (Ayt Ayache)

Language:	Tamazight Berber	*Dialect*:	Ayt Ayache
Phylum:	Afro-Asiatic, Berber, Northern, Atlas		
Inventory:	Kirchner	*Source*:	Abdel-Massih 1971
Lenition(s):	1. b - β (rare)		
	2. k g > x ɣ		
	3. Spirantization blocked in geminates		
Type:	1. Spirantization		
	2. Spirantization		
	3. Distributional property		
Classification:	1. Not Neutralization		
	2. Not Neutralization		
	3. Not Neutralization		

Consonant Inventory

Table 130: Phonemic Consonant Inventory of Ayt Ayache Tamazight Berber

							q		ʔ
		t^h							
b		d							
	f		s	ʃ		x	χ	ħ	h
			z	ʒ		ɣ	ʁ	ʕ	
m		n							
		r							
		l							
w					j				

Description

Process 1 is that in Ayt Ayache the "/b/ which is a voiced bilabial stop can sometimes be heard as voiced bilabial fricative [β] by very few speakers" (p.5).

(3.128) Example of b - β (p.5):

/lbab/ [lβæβ] 'the door'

Process 2 refers to the fact that, although in other dialects of Berber the velar obstruents are stops (k and g), in Ayt Ayache they are fricatives x and ɣ.

What Kirchner lists as process 3 is the observation that spirantization in Ayt Ayache does not, and did not diachronically, affect the geminates.

Discussion

The processes of spirantization (processes 1 & 2) are both not neutralizing. In the case of the bilabial, the β only surfaces in the language as part of the phoneme represented by b, and therefore does not interfere with other phonemes. In the case of the velars, the shift to spirants is complete in this dialect and is no longer an active alternation, so no phonemes are affected by this lenition. Diachronically neutralization was also avoided because the spirants x and ɣ surfaced in the language only as members of the k and g phonemes, and therefore could not have obliterated contrast with other phonemes.

The observation that spirantization does not affect geminates in the language does not have the potential to neutralize contrast because it does not involve an active process.

Additional Comments

It would seem that process 1 and process 2 should be listed together because they affect the same group of sounds (stops b k g) in a similar manner (spirantizing them to β x ɣ). However, the first process is optional and rare, applying to very few speakers, whereas process 2 is probably no longer an active process in the language but instead a diachronic change that has been completed. Therefore they are listed separately.

3.125. Tamazight Berber (Ayt Ndhir)

Language:	Tamazight Berber	*Dialect*:	Ayt Ndhir
Phylum:	Afro-Asiatic, Berber, Northern, Atlas		
Inventory:	Kirchner	*Source*:	Saib 1976
Lenition(s):	1. bb tt dd - β θ ð		
	2. ðˤ zˤ - tˤtˤ sˤsˤ		
Type:	1. Spirantization		
	2. Voicing		
Classification:	1. Not Neutralization		
	2. Not Neutralization		

Consonant Inventory

Table 131: Phonemic Native Consonant Inventory of Ayt Ndhir Tamazight Berber

				tʃ			q	
							qʷ	
				dʒ				
	f	θ	s	ʃ		x	χ	h
						xʷ		
β		ð	z	ʒ		ɣ	ʁ	
		ðˤ	zˤ			ɣʷ		
m		n						
		r						
		l						
w					j			

Description

Process 1 consists of the occlusivization of geminate stops. That is, when β θ ð appear as geminates they also become [-continuant]. Compare (3.129a) where geminated spirants remain continuant with (3.129b, c) where geminated spirants alternate with geminate stops.

(3.129) Geminate stops (p.33):

	zero form	intensive form	
a.	fser	fesser	'to spread'
	rzˤem	rezˤzˤem	'to open'
b.	fθel	fettel	'to fold'
	nðer	nedder	'to roar'
c.	βðˤu	betˤtˤu	'to divide'

Process 2, exemplified in (3.129c), is the voicing of pharyngealized native Berber consonants when they are lax or non-geminate. The fact that this affects native Berber sounds is significant because native pharyngeals are rare. The existence of wide-spread pharyngealization in Northern Berber dialects is not a result of an active native pharyngealization process but is probably an influence from Arabic loans.

Discussion

Saib reports that, with the exception of the nasals and q, all the consonants which occur in non-geminate form are continuants. That is, for native obstruents in the language, with the exception of uvulars, spirantization is not contrastive. To illustrate this point further consider the fact that p t k only surface in the language as spirants (f θ x) or geminates (pp tt kk).[23] With this in mind, the alternation between β θ ð and their [-continuant] geminate counterparts (bb tt dd) cannot obliterate any distinctions in the language because none depend on continuity (or rather, spirantization) for their contrast. The inventory of native phonemes confirms this by listing only the uvular stops, and no other stops in the language, as phonemic.

The voicing of pharyngealized consonants also does not obliterate any contrasts because voicing is not contrastive for pharyngealized native phonemes, as is apparent from the fact that only two such phonemes are pharyngealized, and both are voiced. In other words, if a native sound surfaces as pharyngealized, it must be a representation of one of these two phonemes (ðˤ zˤ); thus, regardless of whether the sound surfaces as voiced or not, it does not neutralize with other existing phonemes in the language.

Additional Comments

Saib lists process 1 as the gemination and subsequent occlusivization of non-uvular spirants (i.e., θ → θθ → tt). Kirchner reanalyzes the process as the degemination and subsequent spirantization (i.e., tt → t → θ). On this topic, Saib reports that "geminates may be exempt from weakening processes (e.g., spirantization) which other consonant sequences undergo" (p.77), adding later that "geminates will reduce (degeminate) first, before undergoing spirantization (or other weakening processes)" (p.78). For this reason Kirchner's analysis of the process as degemination and weakening is preferred to Saib's description of the process as occlusivization.

3.126. Tamazight Berber (Ayt Seghrouchen)

Language:	Tamazight Berber	*Dialect*:	Ayt Seghrouchen
Phylum:	Afro-Asiatic, Berber, Northern, Atlas		
Inventory:	Kirchner	*Source*:	Abdel-Massih 1971
Lenition(s):	b - β (rare)		
Type:	Spirantization		
Classification:	Not Neutralization		

[23] These are the possible sounds in theory, although in the data only tt and dd are attested

Consonant Inventory

Table 132: Phonemic Consonant Inventory of Ayt Seghrouchen Tamazight Berber

							q		ʔ
		t^h				k			
b		d				g			
	f		s	ʃ			χ	ħ	h
			z	ʒ			ʁ	ʕ	
m		n							
		r							
		l							
w					j				

Description

Abdel-Massih reports that in Ayt Seghrouchen, just like in the Ayt Ayache dialect, a few (actually, "very few") speakers sometimes spirantize b to β in all contexts.

Discussion

This process is not neutralizing because β is not phonemic in the language and therefore only surfaces as a member of the b phoneme and does not interfere with other phonemes.

Additional Comments

The spirantization of the bilabial in the Ayt Seghrouchen dialect of Tamazight Berber is listed separately from the same process in the Ayt Ayache dialect (§3.124) mainly due to the fact that the two dialects have different consonant inventories.

3.127. Tamil

Language:	Tamil	*Dialect*:	Not specified
Phylum:	Dravidian, Southern, Tamil-Kannada, Tamil-Kodagu, Tamil-Malayalam, Tamil		
Inventory:	Lavoie	*Source*:	Annamalai 1975
Lenition(s):	p t ts ʈ k → β ð dz ɽ ɣ / [son]_#V in nouns		
Type:	Spirantization and voicing		

Classification: Not Neutralization
Trends: Flapping patterns with spirantization (ʈ - ɽ)
Difference: Omitting one of the (duplicate) Tamils in Kirchner's inventory

Consonant Inventory

Table 133: Phonemic Consonant Inventory of Tamil

p		t	ts	ʈ		k
b		d		ɖ		g
			s	ʂ		
	v					
m		n		ɳ		
		r				
		l		ɭ		
				ɻ	j	

Description

Annamalai reports that the "stops and affricate are voiced and fricativized after a vowel, semivowel or a liquid and before a vowel preceded by a morpheme boundary or an internal word boundary" (p.167-168), and that "if the stop is retroflex in this context, it becomes a voiced retroflex flap" (p.168).

Discussion

The spirantization and voicing of p t ts ʈ k to β ð dz ɽ ɣ is not neutralizing because none of these spirants are phonemic in the language and surface only as members of the phonemes represented by their respective stops. Hence no obliteration of contrast between phonemes is possible.

Notes on Difference from Kirchner's list

In his inventory Kirchner (1998) includes a spirantization process in Tamil from Lavoie's (1996) inventory which is based on Annamalai's (1975) work discussed here, and the same process using a different source (Keating et. al. 1983). I omit the duplicate process.

3.128. Tatar

Language: Tatar *Dialect*: Not specified

Phylum:	Altaic, Turkic, Western, Uralian
Inventory:	Lavoie *Source*: Poppe 1968
Lenition(s):	b d g → β ð ɣ / V_V
Type:	Spirantization
Classification:	Not Neutralization
Difference:	Process description is corrected; duplicate entry is omitted

Consonant Inventory

Table 134: Phonemic Consonant Inventory of Tatar

				tʃ			ʔ
p^h		t^h				k^h	
b		d				g	
	f		s	ʃ	ç	x	h
	v		z	ʒ	ʝ		
m		n				ŋ	
		r					
		l					
w					j		

Description

In medial, mostly intervocalic, context b d g are spirantized. Poppe reports that b and g alternate with their spirant versions β and ɣ, but does not give a clear indication of which sound d alternates with except that it has a weaker articulation with a less complete closure. The alveolar surfaces in its weaker form initially as well as medially.

(3.130) Spirantization in Tatar (p.11):

a.	/bala/	[bala]	'child'
b.	/xäbär/	[xäβär]	'news'
c.	/balaga/	[baɫaɣa]	'to the child'
d.	/kägäz/	[k^häɣäz]	'paper'

Discussion

The spirant versions of b d g surface in the language only as members of these same phonemes. Therefore they do not affect other phonemes in Tatar and cannot be neutralizing.

Notes on Difference from Kirchner's list

Lavoie (1996), cited in Kirchner (1998), lists the process above as context free, but her source (Poppe 1963, of which I use a 2nd revised edition from 1968) specifies that the spirantization occurs mostly in intervocalic context. Lavoie also lists the spirantized version of d as ɾ, but Poppe's description says nothing of flapping the alveolar.[24]

Kirchner includes two entries for Tatar: one from Lavoie's inventory (b d g - β ɾ ɣ), and another that he adds from the same source (Poppe 1963), described as "g → ɣ / mostly V_V". Since the second entry is subsumed by Lavoie's process, it is omitted from my corpus.

3.129. Tauya

Language:	Tauya	*Dialect*:	Not specified
Phylum:	Trans-New Guinea, Madang-Adelbert Range, Adelbert Range, Brahman		
Inventory:	Kirchner	*Source*:	MacDonald 1990
Lenition(s):	1. k kʷ > ʔ ʔʷ non-initially 2. r → t / C_, #_		
Type:	1. Debuccalization 2. Occlusivization		
Classification:	1. Not Neutralization 2. Not Neutralization		
Trends:	Phonetic, but not phonological, neutralizations (both).		
Difference:	Description of the second process is corrected		

[24] In fact, I would not expect the alveolar to flap in Tatar because there is a phonemic trill in this language. In other words, the situation in this language is consistent with my findings regarding the flapping and spirantization of alveolars patterning according to inventory shapes (see §2.3.2.2 for a detailed discussion of this point).

Consonant Inventory

Table 135: Phonemic Consonant Inventory of Tauya

p		t			k	?
					k^w	$?^w$
(b)					(g)	
	f		s			
m		n				
		r				
w				j		

The voiced stops in parentheses occur only in borrowings.

Description

Process 1 is that medially instances of ? and $?^w$ are historically derived from k and k^w (3.131). In fact, except for a few exceptions (3.132c, d) the velars are restricted to word-initial position (3.132a, b). In addition to the diachronic change from velars to glottals, a few forms in the language exhibit synchronic free variation between the velars and glottals in initial position (3.133).

(3.131) Cognate forms in Biyom with medial k (p.24):

	Tauya	Biyom	
a.	[me?eni]	[mekeni]	'coconut'
b.	[nu?upu]	[nukubu]	'lime powder'

(3.132) Distribution of k (p.23):

a.	[kera]	'parrot'
b.	[kau-]	'howl (intransitive)'
c.	[jauke]	'dugong'
d.	[sau eko]	kind of axe

(3.133) Initial free variation k ~ ? (p.24):

a.	[kaifa]	~	[?aifa]	'cockatoo'
b.	[kausi]	~	[?ausi]	place name
c.	[makarafa]	~	[ma?arafa]	personal name

Process 2 is that morpheme-initial r is realized as t in post-consonantal and word-initial positions. This rule applies only to two suffixes: -ra (topic) and -rafo (dubitative) and a handful of stems (e.g., roroti- 'be exhausted (intransitive)').

Discussion

Regarding process 1, note that currently there is no alternation in medial position between the velars and the glottals because, except for very few exceptions, the velars are restricted to initial position. Note also that initially, where velars can surface, velars and glottals indeed belong to different phonemes, as can be seen from the minimal pairs in (3.134). The diachronic origin of the glottals in this context is not discussed in the source.

(3.134) Minimal pairs exhibiting initial contrast (p.23):

a.	[kau-]	'howl (intransitive)'	vs.	[ʔau-]	'swim (intransitive)'
b.	[kera]	'parrot'	vs.	[ʔera]	'that one'

In order to determine whether k k^w > ʔ $ʔ^w$ is neutralizing or not, we need to know whether all the medial glottals in the language are derived from velars. That is, if all instances of medial glottals are derived from velars, there is no neutralization because there is no question as to which phoneme is represented by ʔ and $ʔ^w$ in that context. However, if some of the medial glottals represent the glottal phoneme while others represent the velar phoneme there is definitely neutralization between the glottal and velar phonemes. Or rather, there was neutralization back when the diachronic shift of medial velars to glottals was active in the phonology.

Although MacDonald does not specify the origin of the initial glottals, she does suggest that all instances of medial glottals are derived from velars: she explains that although they may only be allophones of the velars, she assigns the glottals phonemic status based on their word-initial behavior where they do contrast with velars. Hence, based on the glottals' non-initial behavior alone they would not be assigned phonemic status because in that context they are all derived from velars. The answer is therefore that no neutralization resulted from the k k^w > ʔ $ʔ^w$ shift: all instances of glottals in non-initial context represent only the velar phonemes so no contrast between phonemes was obliterated. The phonetic neutralization between the velars and glottals does not result in phonological neutralization.

Process 2 constitutes phonetic neutralization between the r and t phonemes. However, the fact that it affects so few lexical items (two suffixes and a handful of stems) suggests that phonological neutralization is avoided.

Additional Comments

The second process is, according to the author, a case of occlusivization (r → t), which is a type of fortition and not lenition.

Notes on Difference from Kirchner's list

Kirchner (1998) lists the second process for Tauya as t - ɾ, suggesting that t lenites to a flap, a process which is rather common among Kirchner's list of lenitions. His grammar source (MacDonald 1990) however, says nothing of a flap and clearly specifies that the process is one of r → t rather than t → ɾ.

3.130. Thai

Language:	Thai	*Dialect*:	Not specified
Phylum:	Tai-Kadai, Kam-Tai, Be-Tai, Tai-Sek, Tai, Southwestern, East Central, Chiang Saeng		
Inventory:	Kirchner	*Source*:	Noss 1964
Lenition(s):	Final m n ŋ w j are shorter after a long V		
Type:	Shortening		
Classification:	Not Neutralization		

Consonant Inventory

Table 136: Phonemic Consonant Inventory of Thai

p		t		c	k	
p^h		t^h		c^h	k^h	
b		d			g	
	f		s			h
m		n			ŋ	
		r				
		l				
w				j		

Description

The only details Noss provides about the weakening process is that final consonants m n ŋ w j have weaker and shorter allophones after long vowels, and stronger longer allophones after short vowels.

Discussion

Although not much information is provided it is clear that the alternation between the shorter and longer versions of the sonorants m n ŋ w j is not neutralizing because length for these sounds is not contrastive. In other words, regardless of whether they are pronounced as shorter or longer, they clearly represent the same phonemes and cannot be confused with other phonemes in the language.

3.131. Tharu (Chitwan)

Language:	Tharu	*Dialect*: Chitwan
Phylum:	Indo-European, Indo-Iranian, Indo-Aryan, Eastern zone, Unclassified	
Inventory:	Kirchner	*Source*: Leal 1972
Lenition(s):	1. h → Ø / V_V	
	2. p^h b ɖ ɖ^h dʒ - ɸ β ɾ ɾ^h ʒ	
Type:	1. Loss	
	2. Spirantization	
Classification:	1. Incomplete Neutralization	
	2. Not Neutralization	
Trends:	Flapping patterns with spirantization	
Difference:	Four processes are condensed into one (process 2)	

Consonant Inventory

Table 137: Phonemic Consonant Inventory of Chitwan Tharu

p	t		tʃ	ʈ		k	
p^h	t^h		tʃ^h	ʈ^h		k^h	
b	d		dʒ	ɖ		g	
b^h	d^h		dʒ^h	ɖ^h		g^h	
		s					h
m	n					ŋ	
m^h	n^h					ŋ^h	
	r						
	r^h						
	l						
	l^h						
w					j		

Description

In Chitwan Tharu intervocalic h is elided; this is process 1.

(3.135) Examples of h elision (p.24):

a.	/tʰehun/	→	[tʰeun]	'knee'
b.	/sihua/	→	[siua]	'marrow'
c.	/tʰeluha/	→	[tʰelua]	'swing'
d.	/kohar/	→	[koar]	'squash'

Process 2 is the optional spirantization of several stops and affricates in mostly intervocalic context.

(3.136) pʰ ~ ɸ (p.25):

/bʰapʰ/ [bʰapʰ] ~ [bʰaɸ] 'steam'

(3.137) b → β / V_V (p.17):

/debasu/ [deβasu] 'I will give!'

This process also includes the intervocalic and word final flapping of the retroflex voiced stop (3.138a), intervocalic flapping of the aspirated retroflex stop (3.138b), and spirantization of the voiced alveo-palatal affricate (3.139).

(3.138) Flapping of retroflex:

a. ɖ → ɽ / V_V, _#

b. ɖʰ → ɽʰ / V_V

(3.139) Spirantization of alveo-palatal affricate:

dʒ → ʒ / V_V

Discussion

Process 1, the elision of intervocalic h, interacts with another process in the language: the free variation between h and ɦ in various, including intervocalic, contexts. In his transcriptions of the language Leal does not mark either the cases of voiced ɦ or the cases of elided h, although he does give the impression that only the h is elided. In other words, intervocalic h can either voice (to ɦ) or elide (to Ø).

Another relevant detail about Chitwan Tharu is that any cluster of 2 vowels is always disyllabic. Hence, if h were a syllable edge marker in the language (which is a distinct possibility) then its intervocalic elision would not interfere with this function because the resulting vowel cluster would necessary include a syllable boundary.

However, judging by the available information about the language the elision of h does appear to be neutralizing, although how much contrast is actually obliterated is not clear due to the lack of data on the interaction between the loss of h and h ~ ɦ. Therefore, the process is classified as Incomplete-Neutralization.

Process 2, the spirantization of p^h b ɖ $ɖ^h$ dʒ to ɸ β ɾ $ɾ^h$ ʒ is definitely not neutralizing. The resulting spirants and flaps only surface in the language as results of this spirantization process and therefore cannot be confused with existing sounds. Thus no contrast is obliterated.

Notes on Difference from Kirchner's list

In his inventory Kirchner (1998) divides the spirantization process into a list of four, where each affected sound is considered a separate lenition. I consider all four to be part of one process, and list them as such.

3.132. Tigrinya

Language:	Tigrinya	*Dialect*:	Not specified
Phylum:	Afro-Asiatic, Semitic, South, Ethiopian, North		
Inventory:	Kirchner	*Source*:	Kenstowicz 1982
Lenition(s):	1. k q → x χ / V_		
	2. Spirantization blocked in geminates		
Type:	1. Spirantization		
	2. Distributional property		
Classification:	1. Not Neutralization		
	2. Not Neutralization		

Consonant Inventory

Not provided in the source

Description

Post-vocalically k q are spirantized to x χ (process 1). This general rule does not affect geminates (process 2).

(3.140) Spirantization of k q to x χ (p.106):

a.	mə-rkab	(inf.)	räxäbä	(3sg.m. perf.)		'find'
b.	kälbi	'dog'	ʔaxaləb	'dogs'	ʔəti xälbi	'the dog'

Discussion

Kenstowicz reports that "spirantization is not a neutralization rule in Tigrinya. All occurrences of the velar spirants x and χ can be derived from /k/ and /q/ by the spirantization rule" (p.120). Process 1 is therefore not neutralizing.

Process 2 is a distributional observation that geminates are exempt from the spirantization rule listed as process 1. Since this is not an active phonological process it does not have the potential to neutralize any contrasts in the language.

3.133. Toba Batak

Language: Toba Batak *Dialect*: Not specified
Phylum: Austronesian, Malayo-Polynesian, Western Malayo-Polynesian, Sundic, Sumatra, Batak, Southern
Inventory: Kirchner *Source*: Hayes 1986a
Add. Source: Hayes 1986b (Grammar source for Hayes 1986a)
Lenition(s): p t k → ʔ / _C
Type: Debuccalization
Classification: Neutralization
Trends: This neutralizing process helps preserve contrast elsewhere

Consonant Inventory

Table 138: Phonemic Consonant Inventory of Toba Batak (Hayes 1986b)

p	t			k	
b	d		dʒ	g	
		s			h
m	n			ŋ	
	r				
	l				

Hayes notes that the phonemic status of h is unclear.

Description

Hayes (1986a: 340) notes that "the voiceless stops /p t k/ surface as glottal stops whenever they precede a consonant", and calls it the glottal formation rule.

(3.141) Glottal formation (Hayes 1986a: 341):

halak	'person'	halaʔ batak	'Batak person'

Discussion

The pre-consonantal alternation between p t k and ʔ is neutralizing: it obliterates the contrast between the three phonemes p t k. However, according to Hayes 1986b it also helps preserve a different contrast in the language. Data (3.142) shows the interaction between the glottal formation rule and three other processes in Toba Batak (n-assimilation, denasalization, and h-assimilation) and how these affect possible clusters in the language.

(3.142) Rule interaction in Toba Batak (Hayes 1986b: 482):

/pt	pp	np	ŋp	mh	th/	
ʔt	ʔp	--	--	--	--	Glottal Formation
--	--	pp	--	--	--	/n/-Assimilation
--	--	--	kp	kh	--	Denasalization
--	--	--	--	pp	tt	/h/-Assimilation
[ʔt	ʔp	pp	kp	pp	tt]	

The glottal formation rule helps keep clusters such as pp distinct from np and mh, and clusters such as pt distinct from th. It does, however, neutralize the contrast between pp, tp, and kp, all of which result in ʔp.

Additional Comments

The debuccalization rule in Toba Batak is especially interesting because while neutralizing one contrast it also helps maintain another. The question of which contrast is more important in the language, that is, which has the greater functional load, is not one that can be answered based on the available data.

3.134. Tojolabal

Language:	Tojolabal	*Dialect*:	Not specified
Phylum:	Mayan, Kanjobalan-Chujean, Chujean		
Inventory:	Lavoie	*Source*:	Furbee-Losee 1976
Add. Source:	Supple and Douglass 1949		
Lenition(s):	1. r → ɣ / [voiced non-continuant]_V, V_V (optional)		
	2. g → ɣ / V_V (optional)		
	3. h j w → Ø / _#		
Type:	1. Spirantization		
	2. Spirantization		
	3. Loss		
Classification:	1. Not Neutralization		

2. Not Neutralization
3. Neutralization

Trends: Phonetic, but not phonological, neutralization (r/g in processes 1 & 2)

Difference: Processes restated and renumbered

Consonant Inventory

Table 139: Consonant Inventory of Tojolabal (Furbee-Losee 1976)

p	t		tʃ	c	k	ʔ
	$t^{ʔ}$		$tʃ^{ʔ}$	$c^{ʔ}$	$k^{ʔ}$	
(b)	(d)				(g)	
$b^{ʔ}$						
		s	ʃ			h
m	n					
	r					
	l					
w				j		

Sounds in parentheses, b d g, only appear in Spanish loans.

Description

Process 1 is the optional spirantization of r to ɣ between vowels or after a voiced non-continuant and before a vowel. Process 2 is the optional intervocalic spirantization of g.

Process 3 is the word-final loss of h, j and w: h except phrase-finally, j when it follows i, and w when it is post-vocalic.

Discussion

Taken individually, processes 1 and 2 do not lead to neutralization because they result in ɣ, a sound that is not phonemic in the language and therefore does not obliterate contrast with other phonemes. However, since both r and g spirantize to ɣ in the same environment (intervocalically), there is a phonetic neutralization between the r and g phonemes in this context.

Yet the phonetic neutralization between r and g does not result in phonological neutralization: no meaning distinctions are obliterated because contrast between r and g phonemes is retained in the stress pattern of the affected words. Furbee-Losee (1976) reports that Spanish loan words in Tojolabal are treated differently in the phonology, as is evidenced from the fact that these words do not follow the Tojolabal stress pattern. Since g only occurs in Spanish loans, all occurrences of ɣ that represent the g phoneme are in words with a non-native stress pattern. In other words, there is a way to determine which phoneme is represented by an occurrence of intervocalic ɣ: in words with a native stress pattern it represents r, and in words with a non-native stress pattern it represents g. Hence, neutralization between g and r is avoided.

Taken individually, the word final deletions in process 3 (of h j w) may not be neutralizing. In the case of h, there is reason to believe that it is actually predictable at word-boundaries. Glottals are subject to a variety of deletion (such as the word-final deletion of h) and insertion (such as before a word-initial vowel) rules. In fact, Tojolabal has a limited and strict range of possible morpheme shapes (only VC, CVC, CVCC, CVCVC, and CVCCVC are allowed) and it appears that the glottals ʔ and h may be used to shape syllables into allowable patterns.[25]

Also, since all Tojolabal words end in consonants, a word-final vowel must be derived by the deletion of either h, w or j. In the case of the w, its deletion actually maintains contrast with word final Vu sequences: in this environment the vowel u surfaces as w, resulting in a Vw# sequence which is exactly the environment in which phonemic w's are deleted. Furthermore, w is restricted in which vowels it can follow[26], so neutralization with a deleted j may also be limited.

However, regardless of the fact that the individual deletions of h j w may not lead to the obliteration of meaning distinctions, there is neutralization between the results of the deletion of these three sounds. That is, a word-final vowel in Tojolabal surely resulted from the deletion of a consonant, but it is not always clear which of the three sounds was in fact deleted. Therefore, there is neutralization among the h j w phonemes.

---

[25] The suggestion that glottals are word-boundary markers of sorts, and are predictable, is merely that: a suggestion. There is good reason to suppose this but I do not investigate further to prove this hypothesis. The classification of the functional consequences of word-final deletion of h does not depend on this hypothesis and would not change if this hypothesis is wrong.

[26] Supple and Douglass (1949) mention the restriction but do not elaborate on the details. They do mention that w never follows a long i, and that it is one of the few consonants that cannot follow all short vowels, but do not specify which short vowels may or may not precede it.

Notes on Difference from Kirchner's list

Lavoie (1996), cited in Kirchner (1998), lists two processes for Tojolabal which I divide into 4. Her first process is r g → ɣ / V_V, which I divide into r → ɣ and g → ɣ because, according to Lavoie's grammar source (Furbee-Losee 1976) r occurs in an additional environment to intervocalic, and g is a non-native phoneme which is treated differently in Tojolabal phonology.

3.135. Totonac (Misantla)

Language:	Totonac	*Dialect*: Misantla
Phylum:	Totonacan, Totonac	
Inventory:	Lavoie	*Source*: MacKay 1994
Lenition(s):	1. q → χ / V_ (optional) 2. p t ts tʃ k q → b d dz dʒ g ɢ between voiced sounds (optional)	
Type:	1. Spirantization 2. Voicing	
Classification:	1. Not Neutralization 2. Not Neutralization	
Difference:	Contexts for both processes are corrected	

Consonant Inventory

Table 140: Phonemic Consonant Inventory of Misantla Totonac

p	t	ts	tʃ		k	q	ʔ
		s	ʃ				h
m	n						
	l						
	ɬ						
w				j			

Description

Process 1 is the optional spirantization of post vocalic q to χ.

(3.143) q → χ / V_ (p.375):

a.	/ɬuququ-la/	[ɬɔχɔχɔlá]	's/he/it is muddy'
b.	/ʃtuqi/	[ʃtɔ̰́χɛ̰]	'green bean'
c.	/tuq-tuq/	[tɔ́χtɔ́χ]	'robin'

d.	/laqaput/	[laχapút]	'capulin (fruit)'

Process 2 is the voicing of stops and affricates in voiced environments. MacKay writes that "noncontinuant obstruents are optionally voiced between voiced segments. Voicing assimilation occurs much less often between vowels than following a nasal" (p.373).

(3.144) Stop voicing (p.374-375):

a.	/iʃ-tan-pis/	[ʃtámbis]	'her/his navel'
b.	/min-t̰ḭḭh/	[míndɛ̰́ɛ̰h]	'your brother-in-law'
c.	/muu-siink-na̰/	[múusíiŋgna̰]	'caves'
d.	/tʃḭḭ-nḭ-qṵn/	[tʃḭḭɴɢɔ̰́ʔ]	'very fat'

Discussion

Process 1 is not neutralizing because the χ only surfaces in the language as part of its alternation with q, and therefore does not interfere with other phonemes in the language.

Likewise, process 2 is not neutralizing because the voiced stops and affricates only surface in Totonac as members of the phonemes represented by the voiceless stops. The inventory of phonemes shows that voicing is not contrastive for obstruents, so a change in voicing does not obliterate contrast with other phonemes.

Notes on Difference from Kirchner's list

Lavoie (1996), cited in Kirchner (1998), lists the 2 processes in Totonac as word-final spirantization of q and intervocalic voicing of stops. I adhere more closely to Lavoie's grammar source (MacKay 1994) in listing the spirantization of q as occurring post-vocalically, and the voicing affecting affricates as well as stops, and occurring in voiced (but not just intervocalic) contexts.

3.136. Tsou (Tfuya)

Language:	Tsou	*Dialect*:	Tfuya
Phylum:	Austronesian, Formosan, Tsouic, Northern		
Inventory:	Kirchner	*Source*:	Wright 1996
Lenition(s):	ˀɗ → ˀl / _V[low]		
Type:	Flapping & lateralization		
Classification:	Not Neutralization		
Difference:	Process description is corrected		

Consonant Inventory

Table 141: Phonemic Consonant Inventory of Tfuya Tsou

p		t	ts		k	ʔ
ɓ		ɗ				
	f		s			h
	v		z			
m		n			ŋ	
w				j		

Description

The preglottalized dental implosive is realized as a preglottalized lateral flap before a low vowel. Wright provides an example (3.145) but not one that shows an active alternation.

(3.145) Dental implosive and lateral flap (p.33):

a. ^{ʔ}ɗiŋki 'muddy'

b. ^{ʔ}lauja 'maple'

Discussion

The flapping of the dental implosive is not a neutralizing process in Tsou because the lateral flap (be it a lateral l or a flap ɾ) only surfaces in the language as a result of this process and therefore does not interfere with other phonemes.

Notes on Difference from Kirchner's list

Kirchner (1998) lists the lenition process in Tsou as "voiced (implosive) coronal stop lenites to a lateral flap just in case the preceding vowel is low". I adhere more closely to his grammar source (Wright 1996) in describing the process as lateralization/flapping of a preglottalized dental implosive before low vowels.

3.137. Turkana

Language: Turkana *Dialect*: Not specified
Phylum: Nilo-Saharan, Eastern Sudanic, Nilotic, Eastern, Lotuxo-Teso, Teso-Turkana, Turkana
Inventory: Lavoie *Source*: Dimmendaal 1983
Add. Source: Best 1983

Lenition(s):	t → s before a weak morpheme boundary followed by a non-back vowel
Type:	Spirantization
Classification:	Neutralization
Difference:	Process description is corrected

Consonant Inventory

Table 142: Phonemic Consonant Inventory of Turkana (Dimmendaal 1983)

p	t		c	k
b	d		ɟ	g
		s		
m	n		ɲ	ŋ
	r			
	l			
w			j	

Description

In Turkana t → s when preceding a morpheme boundary followed by i, ɪ, e or ɛ (3.146, 3.147). This rule does not apply across word boundaries, or across stronger word-internal boundaries such as those in compounds (3.148a) or in reduplicated forms (3.148b).

(3.146) Examples of t → s (Dimmendaal 1983: 8):

a.	a-kɪ-mat			'to drink'
b.	a-mat-ɛt	→	amasɛt	'cup, instrument for drinking'
c.	a-mat-ɪ	→	amasɪ	'I am drinking'

(3.147) Examples of t → s (Best 1983: 18):

a.	a-ki-lot	'to go'
b.	a-los-i	'I go/shall go'
c.	i-los-ete	'you go/will go'

(3.148) Where t → s does not apply (Dimmendaal 1983: 8):

a.	a-ki-pet	'to kick'
b.	a-pet-e-pet	'to kick repeatedly'

Discussion

This is a clear case of neutralization where the distinction between the t and the s phonemes is obliterated.

Notes on Difference from Kirchner's list

Lavoie (1996), cited in Kirchner (1998), lists the process in Turkana as t → s before non-back vowel, which makes the process seem much more general than it is. I correct the context to more closely reflect the description in Lavoie's grammar source (Dimmendaal 1983).

3.138. Turkish

Language:	Turkish	*Dialect*:	Not specified
Phylum:	Altaic, Turkic, Southern, Turkish		
Inventory:	Lavoie	*Source*:	Underhill 1976
Add. Source:	Bayraktaroğlu and Bayraktaroğlu 1992		
Lenition(s):	1. g → Ø, j, ɣ / V_V (depending on dialect)		
	2. v → β / V_C, v → w / a_u		
Type:	1. Loss (or gliding/spirantization)		
	2. Fronting and gliding		
Classification:	1. Not Neutralization		
	2. Not Neutralization		
Trends:	Phonetic, but not phonological, neutralization (g/Ø)		
Difference:	Process descriptions are corrected; duplicate entry is omitted		

Consonant Inventory

Table 143: Phonemic Consonant Inventory of Turkish (Underhill 1976)

p		t		tʃ	k	
b		d		dʒ	g	
	f		s	ʃ		h
	v		z	ʒ		
m		n				
		r				
		l				

Description

Following vowels, with a few exceptions limited to borrowings, the g is soft (ğ). Process 1 is that between back vowels the soft velar stop is dropped. In some dialects this sound is also dropped between and after front vowels. In other dialects, especially in Anatolia, instead of being lost the ğ surfaces as ɣ, while yet in other dialects it surfaces as j in the same contexts.

Process 2 is the replacement of v with β and w in certain contexts:

> At the beginning of a word, v is pronounced as 'v' in English. In the middle or at the end of a word, the pronunciation of v varies from speaker to speaker, but in almost all cases it is "weaker" than English 'v'. After a vowel and before a consonant, v is frequently pronounced as a bilabial fricative. [...] Between a and u, v may be pronounced as 'w' or may disappear entirely in colloquial speech. (Underhill 1976: 4)

Discussion

Process 1 is not neutralizing. In the dialects where the velar surfaces as a fricative ɣ or glide j the situation is straightforward: neither ɣ nor j are phonemic in the language so no distinctions between phonemes are affected.

Where the velar is lost, although there is a phonetic neutralization with Ø, phonological neutralization is still avoided because the two-vowel sequence resulting from intervocalic loss of the velar is distinguishable from a phonemic long vowel: "A long vowel is a single syllable, pronounced longer than a short vowel; a double vowel is a sequence of two short vowels and is two syllables" (Underhill 1976: 12). Hence, the quantity of the surrounding vowels maintains distinction between the deleted g phoneme and Ø.

Process 2 is also not neutralizing because neither of the sounds that v alternates with (β and w) are phonemic in Turkish, and therefore, in representing the v phoneme do not obliterate distinction with other phonemes.

Notes on Difference from Kirchner's list

Lavoie (1996), cited in Kirchner (1998), lists the two processes for Turkish as (1) Delete or "glide" velar fricative / V_V; and (2) v → β / V_V. I correct both to represent more closely the description in Lavoie's grammar source (Underhill

1976).[27] I also omit Kirchner's second entry for Turkish where he adds a third lenition process for the language: the gliding of v to w, taken from Bayraktaroğlu and Bayraktaroğlu (1992). Since this is subsumed in Lavoie's 2nd process, listing it separately is unnecessary.

3.139. Tzeltal

Language:	Tzeltal	*Dialect*:	Not specified
Phylum:	Mayan, Cholan-Tzeltalan, Tzeltalan		
Inventory:	Lavoie	*Source*:	Kaufman 1971
Lenition(s):	b d g - β ð ɣ		
Type:	Spirantization		
Classification:	Not Neutralization		

Consonant Inventory

Table 144: Phonemic Consonant Inventory of Tzeltal

p		t	ts	tʃ		k	ʔ
$p^{ʔ}$		$t^{ʔ}$	$ts^{ʔ}$	$tʃ^{ʔ}$		$k^{ʔ}$	
b		d				g	
	f		s	ʃ			h
m		n					
		ɾ					
		r					
		l					
w					j		

Description

In Tzeltal the voiced obstruents b d g occur as stops #_V, and C_V, and as spirants β ð ɣ elsewhere.

Discussion

The alternation b d g - β ð ɣ is not neutralizing because the voiced spirants are not part of any phonemes other than the ones represented by b d g, and

[27] Lavoie's second source, Swift 1963, was not available to me; however, both processes Lavoie lists for Turkish were discussed in Underhill 1976 and therefore are included in my inventory.

therefore, in representing the voiced stops do not obliterate distinction with other phonemes.

3.140. Uradhi (all dialects)

Language:	Uradhi	*Dialect*:	Atampaya, Angkamuthi, Yadhaykenu
Phylum:	Australian, Pama-Nyungan, Paman, Northern Pama		
Inventory:	Kirchner	*Source*:	Crowley 1983
Lenition(s):	*p *t *k > β ð ɣ medially		
Type:	Spirantization		
Classification:	Not Neutralization		
Difference:	Process rewritten; citation information corrected		

Consonant Inventory

Table 145: Consonant Inventory of Uradhi

p	t̪	t		c	k
β		ð			ɣ
m	n̪	n		ɲ	ŋ
			ɻ		
		r			
		l			
w				j	

Description

Crowley writes that, for an Australian language, Uradhi is unusual because it has a fricative series (β ð ɣ) contrasting with the stops p t k. The fricative series originated from the lenition of stops in some word-medial environments. First, there was spirantization where the stop followed a word-initial syllable with a long vowel (3.149); note that the vowel was shortened as part of this process.

(3.149) Spirantization after a word-initial syllable with long vowel (p.332):

	Proto-Paman		Atampaya	Angkamuthi	Yadhaykenu	
a.	*pa:ŋkal	>	aɣal	aɣa:	aɣal	'shoulder'
b.	*ka:lka	>	alɣa	alɣa	alɣa	'fall, die'
c.	*pi:ku	>	wi:ɣu	wi:ɣu	wi:ɣu	'rib, side'

Second, there was spirantization following a short vowel in the second syllable of a word:

(3.150) Spirantization after a short vowel (p.332):

	Proto-Paman		Atampaya	Angkamuthi	Yadhaykenu	
a.	*kutaka	>	utaɣa	utaɣa	utaɣa	'dog'
b.	*waɽapa	>	ataɣa	ataɣa	ataɣa	'river'

Discussion

The diachronic spirantization of the stops in Uradhi could not have neutralized any contrasts in the language because we are told that spirants did not already exist in the inventory of the proto language, and in fact, this spirantization process is where the Uradhi β ð ɣ originate. Hence, since the process resulted in sounds that were not already phonemic, no neutralization with existing phonemes was possible.

Notes on Difference from Kirchner's list

Kirchner (1998) lists the process in Uradhi as p t k - β ð ɣ (context unclear). Kirchner's grammar source (Kirchner cites his source as Dixon 1979, but it is Crowley, appearing in Dixon 1983, who describes Uradhi) makes it clear that there is no alternation between stops and spirants in the language. However, he does discuss where the spirants originated as well as the contexts in which they occurred, which is the process I include in my corpus.

3.141. Urdu

Language: Urdu *Dialect*: Not specified
Phylum: Indo-European, Indo-Iranian, Indo-Aryan, Central zone, Western Hindi, Hindustani
Inventory: Kirchner *Source*: Beg 1988
Add. Source: Bokhari 1985
Lenition(s): ɖ $ɖ^h$ ~ ɽ $ɽ^h$ medially and finally
Type: Flapping
Classification: Not Neutralization
Difference: Process description is corrected

Consonant Inventory

Table 146: Consonant Inventory of Urdu (Beg 1988)

p		t			ʈ	c	k	q	
p^h		t^h			ʈ^h	c^h	k^h		
b		d			ɖ	ɟ	g		
b^h		d^h			ɖ^h	ɟ^h	g^h		
	f		s	ʃ			x		h
	v		z	ʒ			ɣ		
m		n					ŋ		
					ɽ				
					ɽ^h				
		r							
		l							
						j			

Description

Beg reports that the retroflex voiced stops ɖ ɖ^h vary with the retroflex flaps ɽ ɽ^h in medial and final positions (3.151), but the stops occur after nasals (3.152). The flaps do not occur initially at all.

(3.151) Retroflex stops and flaps (Beg 1988: 23):

a.	eɽ	~	eɖ	'spur'
b.	bu:ɽ^{h}a:	~	bu:ɖ^{h}a:	'old'

(3.152) Retroflex stops after nasals (Beg 1988: 23):

a.	rã:ɖi:	'small cauldron'
b.	ɖanɖ	'gymnastic exercise'
c.	ɟ^{h}anɖa:	'flag'

A second source on Urdu, Bokhari 1985, provides additional information on the relationship between the retroflex stops and flaps. Bokhari reports that the flaps ɽ ɽ^h are never doubled, and that "they are changed to ɖ ɖ^h for the purposes of doubling" (p.70).

(3.153) Doubling of ɽ ɽ^h (Bokhari 1985: 70):

		from	
a.	paɖɖa	paɽa	'the offspring of a buffalo'
b.	gaɖɖ^{h}a	gaɽ^{h}a	'pit'
c.	buɖɖ^{h}a	bu:ɽ^{h}a	'old man'

Bokhari's only mention of the alternation between the retroflex stops and flaps is his description of denasalization of long vowels preceding a ḍʰ that is changed to ɽʰ.

(3.154) Denasalization before retroflex (Bokhari 1985: 21):

a.	/bā̃ḍʰna:/	baɽʰna:	'to come to an end'
b.	/cā̃ḍʰna:/	caɽʰna:	'to climb, to ride'

Discussion

Neither Beg's description nor his consonant inventory make it clear whether the retroflex stops contrast with the flaps. If they do, the medial and final flapping of the voiced retroflex stops must neutralize this contrast. If they do not, there is no contrast to neutralize.

There are a few helpful facts that Beg provides regarding the retroflex sounds: first, the flaps ɽ ɽʰ only occur in Urdu in medial and final positions, where the alternation takes place. Second, historically, medial and final retroflex stops in Urdu are traced back to retroflex voiced stops in Middle Indo-Aryan (MIA). Since both the Urdu retroflex flap and stop originated from the MIA ḍ, there apparently was no contrast between the two in MIA.

Although it is possible that contrast between the ḍ and ɽ developed in Urdu, there is nothing in either source that suggests this was the case. In listing words that contain the various sounds of the language, for ɽʰ Bokhari (1985: 9) includes caɽʰna: 'to climb', which is a word where the flap was derived from a stop (see 3.154b). While not conclusive proof that no words exist where the flap represents a phoneme other than ḍ's, this does support the view that no contrasts exists between the stops and the flaps. Furthermore, since the flaps only surface where the alternation takes place and nowhere else they are in complementary distribution with the stops, which again supports the no-contrast view. All this, coupled with the fact that historically there was no contrast between the two sounds leads to the conclusion that in Urdu the variation between the retroflex stops and flaps does not obliterate any contrasts.

Additional Comments

Beg's description suggests that the synchronic variation in Urdu between the stops and flaps (ḍ ~ ɽ) is identical to the historical shift from MIA stops to Urdu flaps (MIA ḍ > ɽ Urdu) even as far as context is concerned (medially and finally). That is, the flapping of retroflex stops results in sounds that already exist in the inventory as a result of the exact same process.

Normally, when an alternation results in a sound that already exists in the inventory, and can appear in the same context, contrast is neutralized. However, in this case evidence suggests that, most likely, the existing retroflex flaps do not contrast with the homorganic stops.

Notes on Difference from Kirchner's list

Kirchner (1998) lists the flapping process in Urdu as the alternation between retroflex stops and ɾ. His grammar source (Beg 1988), however, specifies that it is not all retroflex stops but only the voiced counterparts (ɖ ɖ^{h}) that occasionally flap. Furthermore, neither the aspiration nor the place of articulation is affected by this variation (i.e., ɖ ɖ^{h} vary with ɽ ɽ^{h} and not with ɾ).

3.142. Urubu-Kaapor

Language:	Urubu-Kaapor	*Dialect*:	Not specified
Phylum:	Tupi, Tupi-Guarani, Oyampi (VIII)		
Inventory:	Lavoie	*Source*:	Kakumasu 1986
Lenition(s):	p t k → b d g (ɣ) in syllables with non-primary stress (optional)		
Type:	Voicing		
Classification:	Not Neutralization		
Difference:	Process description is restated		

Consonant Inventory

Table 147: Phonemic Consonant Inventory of Urubu-Kaapor

p	t				k	ʔ
					k^{w}	
		s	ʃ			h
m	n				ŋ	
					ŋ^{w}	
	r					
w				j		

Description

The stops p t k are optionally voiced in non-primary stressed syllables. The velar, in addition to voicing, is also spirantized in faster speech.

(3.155) Voicing of p t k (p.399):

a.	arapuha	[arapu'ha]	or [arabu'ha]	'deer'
b.	heta tipe	[he'ta tibe]	or [he'ta dibe]	
			or [he'ta dipe]	'there are many, but in vain'
c.	usak oho	[u'sak oho]	or [u'sag oho]	
			or [u'saɣ oho]	'he went to see'

Discussion

According to the inventory of phonemes, neither voicing nor spirantization are contrastive for obstruents in Urubu-Kaapor. Therefore, the varied pronunciation of p t k does not obliterate distinctions with other phonemes. That is, since b d g and ɣ only represent the p t k phonemes in the language no neutralization is possible.

Notes on Difference from Kirchner's list

Lavoie (1996), cited in Kirchner (1998), lists the process in Urubu-Kaapor as intervocalic optional voicing. Her grammar source (Kakumasu 1986), however, specifies that this process is limited to non-primary stressed syllables and does not list the intervocalic context as part of the requirement for voicing, although all available examples are within this context.

3.143. Uyghur

Language:	Uyghur	*Dialect*:	Not specified
Phylum:	Altaic, Turkic, Eastern		
Inventory:	Kirchner	*Source*:	Hahn 1991
Lenition(s):	b k g q ɢ - ɸ/β xʲ ɣ χ ʁ		
Type:	Spirantization		
Classification:	Not Neutralization		
Difference:	Processes are restated and condensed into one; one is omitted		

Consonant Inventory

Table 148: Phonemic Consonant Inventory of Uyghur

p		t	(ts)	tʃ		k	q	ʔ
b		d		dʒ		g	ɢ	
	f		s	ʃ				
	(v)		z	ʒ		x		h
m		n				ŋ		
		r						
		l						
w					j			

The sounds in parentheses, v and ts, are not phonemic in the standard dialect; ʔ does not occur in codas and ŋ does not occur word-initially.

Description

In Uyghur, all but the coronal stops are spirantized in various environments. The b is spirantized following a vowel and preceding either another vowel or a liquid at the end of a morpheme, in the absence of a pause (3.156). At the end of a morpheme preceding a consonant or a pause the b is spirantized and devoiced to ɸ.

(3.156) Spirantization of b (p.61):

#tøban+# tøβan '(at) the bottom'

The voiced velar g is spirantized and slightly palatalized to ɣ$^{(j)}$ in the coda of a word-initial syllable when not preceding a velar, and intervocalically in the absence of a pause (3.157). The voiceless velar is spirantized and palatalized to x^{j} (3.158c); it has additional variations in different environments (3.158).

(3.157) Spirantization of g (p.65):

a. jiɣda 'an ~ the oleaster'
b. teɣiman 'I (will) reach'

(3.158) Variations of k (p.64-65):

a. Following a consonant, preceding a velar or a pause: k
ikki 'two'

b. Following a front vowel & preceding a pause: k^{j}, k^{jw} if the vowel is rounded
(i) tik^{j} 'steep'

(ii) jyk^{jw} '(the) freight'

c. Following a front vowel and preceding a non-velar consonant in the absence of a following pause: x^{j}, x^{jw} if the vowel is rounded

(i) ax^{j}tip 'active'

(ii) jyx^{jw}ta 'in the freight'

d. Elsewhere: k^{h}

k^{h}ɑlɑ 'bovine animal'

The uvular q is spirantized to χ immediately preceding a non-uvular consonant (3.159a). The voiced uvular ɢ is spirantized to ʁ at the beginning of a word and between two vowels as well as following a voiced consonant other than a non-labial nasal (3.159b).

(3.159) Spirantization of the uvulars (p.67-68):

a. toχsɑn 'ninety'

b. ʁɑz 'a ~ the goose'

Discussion

The variation between the stops b k g q ɢ and their spirantized counterparts ɸ/β x^{j} ɣ χ ʁ does not neutralize distinctions with other phonemes because the spirants only surface in the language as members of the stop phonemes. Note that the only stop of those undergoing spirantization that has a phonemic fricative counterpart is k: x is included in the inventory of Uyghur phonemes. However, the spirantized variation of k results in a palatalized fricative (x^{j}) which is distinct from the phonemic x, and hence neutralization between the k and x phonemes is avoided. None of the other spirants (ɸ/β ɣ χ ʁ) are phonemic in the language.

Notes on Difference from Kirchner's list

Kirchner (1998) lists 6 lenitions in Uyghur: (1) b → β / V_ {V, liquid}; (2) k, g → x / V_C; (3) g → ɣ / in coda of wd-initial syllable and /V_V; (4) q → χ in coda; (5) ɢ → ʁ / #_ and V_V, ɢ → χ / _-voi C; and (6) all process blocked in full or partial geminates. I condense the first 5 items into one: the spirantization of non-coronal stops. I also correct the contexts of these processes to more closely reflect the description in Kirchner's grammar source (Hahn 1991). The 6th entry in Kirchner's list, the blocking of spirantization in geminates, could not be found in the source and is therefore omitted.

3.144. Uzbek

Language:	Uzbek	*Dialect*:	Standard
Phylum:	Altaic, Turkic, Eastern		
Inventory:	Kirchner	*Source*:	Sjoberg 1963
Lenition(s):	b p q ~ w ɸ x/ɣ		
Type:	Spirantization		
Classification:	Neutralization		
Difference:	Processes are combined into one		

Consonant Inventory

Table 149: Phonemic Consonant Inventory of Uzbek

p		t	(ts)	tʃ		k	q	
b		d		dʒ		g		
ɸ	(f)		s	ʃ		x		h
			z	(ʒ)		ɣ		
m		n				ŋ		
		r						
		l						
w					j			

Sounds in parentheses are rare.

Description

Sjoberg reports that in Uzbek there is free variation between the following stops and spirants: p ~ ɸ especially in medial position (3.160a), b ~ w in medial position (3.160b), q ~ x medially or finally (3.160c), and q ~ ɣ in final position (3.160d).

(3.160) Free variation between stops and spirants (p.19):

a.	ittipɔq	~	ittiɸɔq	'union'
b.	kabɔb	~	kawɔb	'shashlik'
c.	toqsɔn	~	toxsɔn	'ninety'
	issiq	~	issix	'hot; heat'
d.	pitʃɔq	~	pitʃɔɣ	'knife'

Discussion

Based completely on Sjoberg's list of phonemes in Uzbek, all the sounds involved in the process (p b q and ɸ w x ɣ) are phonemic. Thus, the spirantization of p b q neutralizes the distinction between the stop and spirant phonemes. The extent of this neutralization is somewhat reduced by the fact that the sounds are in free variation.

Additional Comments

Sjoberg refers to the variations described here as "alternations among phonemes" (p.19) and formalizes his examples as such (e.g., 3.160a is written as /ittipɔq ~ ittiɸɔq/ 'union'). In other words, he does not seem to claim that the phoneme p is realized as a spirant ɸ, but instead, that the phoneme p is sometimes replaced by the phoneme ɸ. If this were the case, and a strange one it would be, there would be no neutralization between phonemes because each sound would represent its own phoneme. Loss of meaning distinction would result from the ɸ representing the p phoneme in a context where it can also represent the ɸ phoneme, not from a situation where each sound represents its own phoneme.

Most likely Sjoberg uses the term *phoneme* differently than it is commonly used in literature, in which case the inventory of what he refers to as *phonemes* in the language may very well be a list of all sounds. If this is the case, the classification of the spirantization of p b q as neutralizing is incorrect. No other sources on Uzbek were available, so for the purpose of the present study I classify the spirantization in Uzbek as neutralizing with this discussion as a warning to the reader that the classification may very well be based on faulty description.

Notes on Difference from Kirchner's list

Kirchner (1998) lists the spirantization of p b q in Uzbek as 3 separate processes. I condense the spirantization of the three similar sounds, in similar environments, into one process.

3.145. Vietnamese (Vinh)

Language:	Vietnamese	*Dialect*: Vinh
Phylum:	Austro-Asiatic, Mon-Khmer, Viet-Muong, Vietnamese	
Inventory:	Kirchner	*Source*: Emeneau 1951

Lenition(s): ɣ - g
Type: Occlusivization
Classification: Not Neutralization
Difference: Context is corrected

Consonant Inventory

Table 150: Phonemic Consonant Inventory of Vinh Vietnamese

		t			c	k	
		t^h					
				$ʈ^{r}$		k^x	
b		d					
	f		s	ʂ			h
	v		z	ʐ		ɣ	
m		n			ɲ	ŋ	
		l					

The above are listed by Emeneau as "initial phonemes"; in final position t t^h and ʈ "are in complementary distribution and are to be classed as allophones of one phoneme" (p.13), as are the nasals (m n ɲ ŋ).

Description

Emeneau reports that "the initial voiced velar stop g and the initial voiced velar spirant ɣ are in complementary distribution and are to be classed as allophones of one phoneme ɣ" (p.12). No context is provided.

Discussion

Since only one of the sounds participating in the g - ɣ alternation is listed as phonemic in the inventory (the spirant ɣ), the alternation does not involve sounds that represent other phonemes and therefore no neutralization is possible.

Additional Comments

Emeneau lists the spirant as the basic phoneme, suggesting that the alternation is not one of spirantization (g → ɣ) but one of occlusivization (ɣ → g). If this is the case then this is fortition rather than lenition.

Notes on Difference from Kirchner's list

Kirchner (1998) lists the lenition process in Vietnamese as "g → ɣ except when preceding word ends in ŋ". However, Kirchner's grammar source (Emeneau 1951) mentions no context for this process which is why this part of Kirchner's description is omitted.

3.146. Warndarang

Language:	Warndarang	*Dialect*:	Not specified
Phylum:	Australian, Maran, Mara		
Inventory:	Kirchner	*Source*:	Heath 1980
Lenition(s):	1. p t ʈ c k → b d ɖ ɟ g except syllable-finally 2. p c → w j / V_V in reduplications (rare, with exceptions)		
Type:	1. Voicing 2. Gliding		
Classification:	1. Not Neutralization 2. Not Neutralization		
Trends:	Phonetic, but not phonological, neutralization (process 2)		
Difference:	Process descriptions are corrected		

Consonant Inventory

Table 151: Phonemic Consonant Inventory of Warndarang

p	t̪	t	ʈ	c	k
m		n	ɳ	ɲ	ŋ
			ɽ		
		r			
		l	ɭ		
w				j	

The dental t̪ is uncommon and is mostly found in loans.

Description

Process 1 is the voicing of p t ʈ c k to b d ɖ ɟ g in almost all environments except syllable-finally. Note that this does not apply to t̪, which is a rare non-native phoneme. The voicing process is an integral part of the following process.

Process 2 involves certain rare reduplications: stem-initial p and c in the reduplicant, in intervocalic context, surface as glides w and j (3.161), but with exceptions (3.162).

(3.161) The gliding of p c to w j (p.12):

a.	ɟaɽi-jaɽi	'to do continuously'
b.	biji-wijima	'(reciprocal aux v.)'

(3.162) Exception to the gliding rule (p.13):

buwa-buwa (not buwa-wuwa) 'to face punishment by spearing'

Discussion

Process 1 is not neutralizing: the inventory of phonemes shows that voicing is not contrastive and that the voiced and voiceless stops are members of the same phonemes.

Process 2 appears to be neutralizing because the glides w j that alternate with the stops p c are phonemic in Warndarang, and furthermore, can occur in intervocalic context where this alternation applies. However, since the gliding of p c occurs only in reduplications, and applies only to the reduplicant, there is no chance of confusing the glides involved in this alternation with existing (phonemic) glides in the language. That is, the glides that belong to the p c phonemes constitute the only differences between the reduplicant and the base in the forms where the alternation occurs, and therefore, are recognizable as the p c phonemes. Neutralization between the stop and the glide phonemes is avoided.

Additional Comments

Process 2 involves phonemic overlap since the w j can be members of the glide phonemes (w j) as well as the stop phonemes (p c).

Notes on Difference from Kirchner's list

Kirchner (1998) lists the processes in Warndarang as $T^j \rightarrow j$ / V_V and K P → w / V_V. His use of capitals is meant to include the voicing process as part of the gliding. I list the general voicing process in Warndarang as a separate process and condense the gliding processes into a second item. The context for gliding is corrected to reflect the description in Kirchner's grammar source (Heath 1980), which specifies that this only occurs in reduplications. Finally, the gliding of the velar is omitted because it is limited to only one case and therefore should not be considered a general process in the language.

3.147. West Tarangan

Language:	West Tarangan	*Dialect*:	Not specified
Phylum:	Austronesian, Malayo-Polynesian, Central-Eastern, Central Malayo-Polynesian, Aru		
Inventory:	Lavoie	*Source*:	Nivens 1992
Lenition(s):	1. j w → g dʒ / #_ and medially in unstressed syllables 2. k ~ ʔ / V_V, where both vowels are non-high		
Type:	1. Gliding (or rather de-gliding) 2. Debuccalization		
Classification:	1. Not Neutralization 2. Not Neutralization		

Consonant Inventory

Table 152: Phonemic Consonant Inventory of West Tarangan

	t			k
b	d			
ɸ		s		
m	n			ŋ
	r			
	l			
w			j	

Description

The first process is described by Nivens as the "glide-to-obstruent" rule: "both word initially and in the onset of a stressed non-initial syllable, the glides /w/ and /j/ are realized as [g] and [dʒ] respectively" (p.139).

(3.163) De-gliding of w and j (p.139):

a.	wur	[ˈgur]	'grass'
b.	wɔwa	[ˈgɔwə]	'blossom'
c.	jɔm	[ˈdʒɔm]	'grandparent/grandchild'
d.	bijarum	[biˈdʒarum]	'whale'

The second process is described by Nivens as "k-weakening": "in fast or casual speech /k/ may become glottal stop between two non-high vowels. This cannot happen at a word boundary, although it may occur at a clitic boundary" (p.146).

(3.164) Debuccalization of k (p.146):

a.	gaka	['gakə] ~ ['gaʔa]	'swamp'
b.	ɔk-kanaŋ	[ɔkə'naŋ] ~ [ɔʔə'naŋ]	'my, mine'

Discussion

Process 1 is not neutralizing because g and dʒ only surface in the language as outcomes of the de-gliding of w and j and therefore do not interfere with existing phonemes.

Process 2 is also not neutralizing for the same reason: ʔ only surfaces in the language as a result of its alternation with the k phoneme, and does not interfere with other phonemes.

Additional Comments

The directionality of process 1 (w j - g dʒ) is specifically addressed by Nivens, who claims the process is one of occlusivization (normally considered fortition) rather than gliding (normally considered lenition).

3.148. Wiyot

Language:	Wiyot	*Dialect*:	Not specified
Phylum:	Algic, Wiyot		
Inventory:	Lavoie	*Source*:	Teeter 1964
Lenition(s):	No word-final aspirated sounds		
Type:	Deaspiration / Distributional property		
Classification:	Not Neutralizing		

Consonant Inventory

Table 153: Phonemic Consonant Inventory of Wiyot

p	t	ts	tʃ		k	
p^{h}	t^{h}	ts^{h}	tʃ^{h}		k^{h}	
					k^{w}	
					k^{wh}	
b					g	
		s	ʃ			h
m	n					
	ɾ					
	l					
	ɬ					
w	ɹ			j		

Teeter lists d as a phoneme, adding that it is always pronounced as a flap; since it is a flap in every environment I list the phoneme as ɾ.

Description

The only mention of what Lavoie refers to as word-final deaspiration of stops is Teeter's comment that while plain stops are unrestricted with respect to their distribution in a word, aspirated stops may not appear in final position. Teeter adds that the aspirated stops are also less common than the plain ones.

Discussion

The fact that aspirated stops are limited in their distribution does not necessarily imply that there is, or was, word-final deaspiration in Wiyot, and according to the grammar there does not seem to be a phonological process that deaspirates stops. More importantly for our purposes, the distributional property is not an active process that leads to a sound change and therefore does not have the potential to lead to neutralization.

3.149. Yakut

Language:	Yakut	*Dialect*:	Not specified
Phylum:	Altaic, Turkic, Northern		
Inventory:	Kirchner	*Source*:	Krueger 1962
Add. Source:	Barashkov 1953 (Krueger's grammar source)		

Lenition(s):	1. p → b / V_V
	2. k → χ / adjacent to non-high vowels
	3. No voiced geminate obstruents
Type:	1. Voicing
	2. Spirantization
	3. Distributional property
Classification:	1. Not Neutralization
	2. Not Neutralization
	3. Not Neutralization
Trends:	Process 1 involves some phonetic, but not phonological, neutralization
Difference:	One process is omitted

Consonant Inventory

Table 154: Consonant Inventory of Yakut (Krueger 1962)

(p)	t		tʃ		k		
b	d		dʒ		g		
		s				(χ)	h
						ʁ	
m	n			ɲ	ŋ		
	r						
	l			ʎ			
				j			

Sounds in parentheses: χ exists in the language only as an allophone of k; p is rare and mostly non-native.

Description

Process 1 is described by Krueger as the devoicing of word-final b, adding that "finally, p occurs only as a morphemic /b/ devoiced to p" (p.56). A more traditional formalization of what is exemplified in (3.165) is, as described by Kirchner, the intervocalic voicing of p to b.

(3.165) Alternation between p and b (p.56):

a.	saba	'his thread'	sap	'thread'
b.	tabar	'he hits'	tap-	'to hit'

Process 2 is the alternation between the velar stop k and the uvular spirant χ: "/k/ is used after /ɨ i u y/ and before /e ø i ɨ u y ɨa ie uo yø/; /χ/ is used before /a o/ and after /a e o ø ɨa ie uo yø/" (Krueger 1962: 59).

Process 3 is the observation that voiced obstruents do not contrast in length: "The long consonants in Yakut are /p m t s ŋ ɲ n l tʃ dʒ k χ/ (see Barashkov 1953: 11). Therefore those that are not long are /b j r g ʁ d/" (Krueger 1962: 54).

Discussion

Process 1 is described as word-final devoicing by Krueger, and intervocalic voicing by Kirchner. Since upon prefixation the stem-final p is voiced, Kirchner's analysis is preferred. In Yakut p is a member of two different phonemes: the p phoneme that is described as marginal, rare, and mostly non-native, and the p/b phoneme. The question is whether there is neutralization between these two phonemes, and the answer is simple: Krueger specifies that the only word-final p in the language is the one that belongs to the p/b phoneme, and that is also the only context where this p can surface. Hence, neutralization between the p and the p/b phonemes is avoided because the two p's never occur in the same context.

Krueger makes it clear that the two alternating sounds in process 2, k and χ, belong to the same phoneme. He lists χ as separate only "for instructional purposes... to retain a transcription as close as possible to the official orthography" (p.60). In other words, there is no neutralization between phonemes because both sounds are members of only one and the same phoneme.

The third item on Kirchner's list for Yakut does not involve an active process in the phonology, but an observation that voiceless obstruents do not contrast in length. This does not have the potential to neutralize contrast because it does not involve any variations in how phonemes are pronounced.

Additional Comments

The first process, the alternation p - b, involves a phonetic neutralization between the p and p/b phonemes. However, as discussed above, it does not lead to phonological neutralization because the two p's belonging to separate phonemes cannot occur in the same context.

Notes on Difference from Kirchner's list

Kirchner (1998) adds one more process to his list of lenitions in Yakut: t → d / V_#V. Kirchner's grammar source (Krueger 1962), however, makes no mention of the voicing of t. In fact, his only discussion of the alveolar stops t and d centers on the contrast between them. Therefore, this process is omitted.

3.150. Yana

Language:	Yana	*Dialect*:	Not specified
Phylum:	Hokan, Northern, Yana		
Inventory:	Lavoie	*Source*:	Sapir & Swadesh 1960
Lenition(s):	1. p t ts k partially voice _V 2. b → w / V_V		
Type:	1. Voicing 2. Gliding		
Classification:	1. Not Neutralization 2. Limited Neutralization		
Trends:	Process 2 is phonetic, but only limited phonological, neutralization		

Consonant Inventory

Table 155: Consonant Inventory of Yana

p	t	ts		k	ʔ
p^h	t^h	ts^h		k^h	
$p^ʔ$	$t^ʔ$	$ts^ʔ$		$k^ʔ$	
		s		x	h
m	n				
	r				
	l				
w			j		

Description

Process 1 is described as: "the pronunciation of /p t ts k/ evidently varies between fully voiceless in preconsonantic position and partly voiced in prevocalic positions" (p.3).

Process 2 is the fact that initial b, when it finds itself in post-initial position, is replaced by w (3.166). This is limited to a few b-initial stems, but not all.

(3.166) b - w (p.7):

bal(-la) -wal(-la) 'mouth'

Discussion

Process 1 is not neutralizing: there is only one set of simple (i.e., not aspirated or glottalized) stops in the language's phoneme inventory, which vary in their pronunciation from voiceless to partially voiced. They do not interfere with the pronunciation of any other phonemes, and therefore do not lead to obliteration of any contrasts.

Process 2 does result in neutralization: since phonemic w can occur stem-initially, the alternation between stem-initial b and w neutralizes the distinction between the b and w phonemes. However, this is limited to only a small number of stems (out of a large number of possible forms that start with b) and therefore does not have a sweeping affect. In fact, it is doubtful any meaning distinctions are obliterated by this process due to the limited number of forms it affects. I classify it as limited (phonological) neutralization. Note that b - w does constitute phonetic neutralization.

3.151. Yindjibarndi

Language:	Yindjibarndi *Dialect*: Not specified
Phylum:	Australian, Pama-Nyungan, South-West, Coastal Ngayarda
Inventory:	Kirchner *Source*: Wordick 1982
Add. Source:	O'Grady 1966 (Wordick's source for proto-language data)
Lenition(s):	1. p t̪ ɟ k → w j j w / V_V
	2. ʈ → ɽ / V_V
Type:	1. Gliding
	2. Flapping
Classification:	1. Not Neutralization
	2. Not Neutralization
Trends:	Apparent phonetic, but not phonological, neutralization (Process 1)
Difference:	Process descriptions are corrected and regrouped; some are omitted

Consonant Inventory

Table 156: Consonant Inventory of Yindjibarndi (Wordick 1982)

p	t̪	t	ʈ	c	k
m	n̪	n	ɳ	ɲ	ŋ
		ɾ			
		l	ɭ		
w	j̪		ɻ	j	

j̪ is a j with an "advanced" diacritic meant to represent an interdental glide that is a fronted j.

Description

Process 1 is the intervocalic gliding of the following stops: p → w / V_V in reduplications (3.167a); t̪ → j̪ / V_V in reduplications (3.167b); ɻ → j / a_i (3.167c); and k → w / u_ (3.167d).

(3.167) Gliding of stops:

a.	paɻpa- + reduplicative -paɻpa		paɻpawaɻpa	'thumpity-thump' (p.28)
b.	t̪aɾu- + reduplicative -t̪aɾu		t̪aɾuj̪aɾu	'obscured'(p.28)
c.	maɻkuɾa 'good' + -ɻi-		maɻkuɾaɻi- maɻkuɾaji[28]	'come good' (p.30)
d.	paɭtu 'feather' +	-kaɭlaa 'having'	paɭtuwaɭlaa 'bird'	
	ŋawuɾ 'bubble' +	-kaɭlaa 'having'	ŋawuɾaɭlaa	'beer' (p.28)

Process 2 is the intervocalic flapping of ʈ to ɽ; no examples are provided.

Discussion

Process 1, the (mostly) intervocalic gliding of some stops, is not neutralizing despite the fact that it appears to be so because glides are phonemic in Yindjibarndi, and because both p and k alternate with w. Neutralization with existing glides is avoided because intervocalic glides in the language were diachronically deleted. Thus, intervocalic gliding of the stops p t̪ ɻ k results in the only glides that can surface intervocalically. As for the possible neutralization between the result of gliding p and k, that too is avoided because p - w (as well as t̪ - j̪) is limited to reduplications, where the phonemic identity of the glide is obvious from the stop it "duplicates" in the base.

[28] Only one informant ever used this form.

Process 2, the intervocalic flapping of ʈ, is not neutralizing because the retroflex flap is not a member of any other phonemes in the language and therefore no obliteration of contrast between phonemes results from this alternation.

Notes on Difference from Kirchner's list

Kirchner (1998) lists 9 lenition processes in Yindjibarndi: (1) p → w / V_V; (2) k → w / u_; (3) k → Ø / V_V; (4) t̪ → ð̪ / V_V; (5) tʲ → j / V_V; (6) glides → Ø / V_V; (7) t ʈ - ɾ; (8) r → j / a_i; and (9) r → Ø / V_V. These are divided into 4 general categories: (1) the gliding of intervocalic stops, listed as process 1; (2) the flapping of ʈ, listed as process 2; (3) diachronic processes (e.g., loss of intervocalic glides, and the flapping of t) which are omitted because no contrastive information exists in the sources that helps determine the classification of these processes as neutralizing or not; and (4) processes that are not found in Kirchner's grammar source (Wordick 1982), e.g., voicing of t̪, which are also omitted.

3.152. Yolngu (Djapu)

Language: Yolngu *Dialect*: Djapu
Phylum: Australian, Pama-Nyungan
Inventory: Kirchner *Source*: Murphy 1983
Lenition(s): t̪ c ~ j, p k ~ w
Type: Gliding
Classification: Neutralization
Difference: The two gliding processes are condensed into one

Consonant Inventory

Table 157: Consonant Inventory of Djapu Yolngu

p	t̪	t	ʈ	c	k
			ɖ		
m	n̪	n	ɳ	ɲ	ŋ
		r			
		l	ɭ		
w			ɻ	j	

Description

In Djapu Yolngu there are semivowel alternants in word-medial position following a vowel, liquid, or semivowel: t̪ and c alternate with j (3.168a, b), and p and k alternate with w (3.168c, d).

(3.168) Examples of gliding (p.29):

a. minjʔci 'colour, paint' + d̪arpu-NG 'pierce'
   minjʔci-jarpu-NG 'paint'

b. ɟarakaɭaʔju-N 'move in an uncontrolled way'
   ɟarakaɭaʔ-ɟarakaɭaju-N 'keep moving in an uncontrolled way'

c. d̪a: 'mouth' + birkaʔju-N 'try'
   d̪a:-wirkaʔju-N 'ask'

d. d̪awal 'country' + gujaŋi-Ø 'I think'
   d̪awal-wujaŋi- Ø 'be born'

Discussion

The gliding of t̪ c and p k obviously leads to neutralization between the place of articulation of the two sounds undergoing gliding in each case. That is, there is neutralization between the dental and the palatal stops because both surface as j, and neutralization between the labial and velar stops because both surface as w.

Additional Comments

The classification of the gliding in this language as neutralizing is rather straightforward, but the situation around it is quite interesting. As it turns out, there was a similar diachronic gliding process in the language that affected the voiced counterparts of the sounds discussed here (*ɟ *d̪ > j, *b *g > w). The author argues that the lenis/fortis contrast (in his opinion manifested as the voiced/voiceless contrast) was eliminated in the language by the gliding of the lenis counterparts in all but the retroflex stops. For the current situation, the author reports that both voiced and voiceless stops occur in the language but they are in complementary distribution and do not contrast phonemically. In fact, current gliding occurs where t̪ c p k surface as voiced. So on the surface, the current gliding in the language affects the same sounds as did the diachronic gliding (d̪ ɟ > j, b g > w) although the current phonemes are actually the original voiceless series.

Since the glides j and w are reconstructed from *ɟ *d̪ and *b *g, the contrast between the stops and the glides (between t̪ c and j and between p k and w) is the present-day manifestation of the voicing contrast in the proto language (i.e., between t̪ c and d̪ ɟ and between p k and b g). Thus, not only does the present-day gliding neutralize the place of articulation between dentals and palatals and between labials and velars, but it also neutralizes the original contrast between the lenis/fortis stops of the proto-language.

Notes on Difference from Kirchner's list

Kirchner (1998) lists the gliding in this language as two processes: laminal stops (dental t̪ and palatal c) → ð, j, and k, p → w/{V, liquid}_ (w/ lexical exceptions). I follow Kirchner's grammar source (Murphy 1983) by rewriting the two processes as part of the same general phenomenon: the gliding of stops. I also omit t̪ - ð, which is not found in Murphy's work.

3.153. Yuman

Language:	Yuman	*Dialect*:	Yuman is a language family
Phylum:	Hokan, Esselen-Yuman, Yuman		
Inventory:	Lavoie	*Source*:	Wares 1968
Lenition(s):	c > s (context unclear)		
Type:	Spirantization (or fronting)		
Classification:	Not Neutralization		
Trends:	In Cocopa, diachronic phonemic overlap *c > s, *s > ʂ		
Difference:	One process is omitted		

Consonant Inventory

Yuman is a family of languages, and the author discusses the following 9: Walapai (Northern Yuman: NYu), Mohave (Mo), Maricopa (Ma), Yuma (Yu), Cocopa (Co), Diegueño (D), Tipai (T), Paipai (P), & Kiliwa (K). No individual inventories are provided.

Description

For the 9 languages listed above the author provides the following reconstructions of the proto *c[29], *s, *ʂ.

[29] In this particular case there is not enough information to determine whether the author means the palatal c or the affricate ts by his use of the symbol 'c'. Contrast decisions are not affected by either choice.

Table 158: Reconstruction of Proto-Yuman Sibilants

PYu	NYu	Mo	Ma	Yu	Co	D	T	P	K
*c	c	c	c	c	s	c	c	c	c
*s	θ	θ	s	s	ʂ	s	s	s	s
*ʂ	s	s	ʃ	ʃ	ʃ	ʃ	ʃ	ʂ	s

Discussion

The diachronic shift of *c > s is only attested in one language (Cocopa) in which *s > ʂ and *ʂ > ʃ, and ʃ did not exist in the proto language. Thus, the shift preserved the contrast between c, s, and ʂ, and the spirantization of c did not lead to neutralization.

Notes on Difference from Kirchner's list

Lavoie (1996), cited in Kirchner (1998), adds *v > w to her processes in Yuman. This process clearly was part of a merger, however, there is not enough information to determine whether it resulted in the neutralization of any contrasts. Therefore it is omitted from my inventory.

CHAPTER 4

Implications

In this chapter I summarize the trends that emerge in my corpus of (mostly) lenition processes and express these as general tendencies of such phenomena. The more significant of these trends are:

1. *Lenition processes are overwhelmingly meaning-maintaining*: 92% of such sound changes in the corpus avoid phonological neutralization; even in the case of loss 71% of the processes are non-neutralizing. Furthermore, at almost ¼ of the corpus, 52 processes result in phonetic, but not phonological, neutralization.
2. *Syllable context is significant*: processes that are neutralizing are more likely to affect the pre-consonantal position, while processes that affect the pre-vocalic position are less likely to obliterate meaning distinctions.
3. *Contrast maintenance strategies*: existing contrasts in a language affect the progress and outcome of phonetically-conditioned processes. Voicing is 79% more common where this feature is not contrastive, and spirantization is 92% more common under the same circumstances. The outcome of the spirantization of alveolar stops depends on the shape of the phonemic inventory in a given language. And the more common distinction maintaining strategies observed in cases of phonetic neutralization and sound mergers involve system-wide changes.

Phonetically conditioned sound changes such as lenition are attributed to biological factors which include the physical shape of the vocal organs, their movement, and the acoustic consequences of these properties. As such, these sound changes are motivated by properties that are common to all speakers regardless of their language, and are divorced from all phonological considerations. However, while such changes are motivated by physical properties of the articulators, the present study strongly suggests that the progress and outcome of these changes is in fact constrained by the functional consideration of contrast-maintenance.

The fact that so many of the (mostly) lenition processes investigated in the present study are found to be non-neutralizing overwhelmingly supports the hypothesis that such processes do not operate independently of functional considerations. The present study goes further into the nature of the relationship between phonetic changes and functional considerations, allowing a glimpse into the progress and outcome of the sound changes under circumstances where contrast is threatened to various degrees.

Sound changes that do not threaten contrast are far more likely to proceed unhindered, with widespread consequences as in the case of most voicing and spirantization in the corpus where the changes affect entire series of sounds (e.g., the entire stop series in a language). The progress of sound changes that somewhat threaten contrast is directly influenced by the system of contrasts in a language: the outcome of the spirantization of alveolar stops in the corpus is shown to strongly pattern according to the shape of the phonemic inventory in a language, where flapping is avoided only in languages where r is phonemic. And sound changes that threaten contrast, as in the case of sound mergers in the corpus, often induce further changes that reshape phonological systems and help maintain contrast.

Within the corpus neutralization is more common in non pre-vocalic position where contrast is less perceptible and more difficult to maintain (Steriade 1999). This further suggests that the relationship between phonetically conditioned changes and contrast is not arbitrary.

Of the four contrast maintaining strategies observed in non-neutralizing sound mergers in the corpus, which are low functional load, distributional properties, phonemic overlap and contrast shifts, the last two affect entire systems in the given languages and are also the most common. Such system-wide changes would not be so common if contrast played the role of a 'filter', simply blocking sound changes that threaten to obliterate meaning distinction. Instead, it appears that systems of contrast in languages exert a gradual diachronic force over phonetic processes, a force which plays a prominent role in shaping phonological systems.

The predictive powers of this view are not deterministic, but rather probabilistic. The present study predicts that lenition processes will spare contrast about 90% of the time. Furthermore, it predicts that if contrast is not threatened at all the sound change will be widespread, affecting entire series of sounds. On the other hand if contrast is threatened, especially in the case of oppositions with high functional loads, it will most likely be preserved by way of systemic changes in the language.

Bibliography

Abbott, Miriam. 1991. *Macushi*. In Derbyshire, Desmond C. and Geoffrey K. Pullum (Eds.). "Handbook of Amazonian Languages, Volume 3". 23-160.

Abdel-Massih, Ernest T. 1971. *A Reference Grammar of Tamazight: A Comparative Study of the Berber Dialects of Ayt Ayache and Ayt Seghrouchen*. Center for Near Eastern and North African Studies, The University of Michigan: Ann Arbor, Michigan.

Abraham, P. T. 1985. *Apatani Grammar*. Central Institute of Indian Languages Grammar Series-12. E. Annamalai (Ed.). D. P. Pattanayak, Mysore-570006.

Acharya, Jayaraj. 1991. *A Descriptive Grammar of Nepali and an Analyzed Corpus*. Georgetown University Press: Washington, D.C.

Acharya, K. P. 1983. *Lotha Grammar*. Central Institute of Indian Languages Grammar Series-10. E. Annamalai (Ed.). Manasagangotri, Mysore-570006.

Alderete, John and Tanya Bob. 1999. *An Annotated Bibliography of Tahltan Language Materials*. MS., Rutgers University.

Alderete, John. 2001. *On Tone and Length in Tahltan*. John Benjamins.

Anderson, S. 1981. "Why Phonology Isn't Natural". *Linguistic Inquiry* 12:4: 493-539.

Andrésen, Bjørn Stålhane. 1968. *Pre-glottalization in English Standard Pronunciation*. Norwegian Universities Press: Oslo.

Andronov, M. S. 1969. *The Kannada Language*. Nauka Publishing House, Central Department of Oriental Literature: Moscow.

Annamalai, E. 1975. *Phonology of Tamil Nouns*. In Schiffman, Harold F. and Carol M. Eastman (Eds.). "Dravidian Phonological Systems". 123-179.

Aoki, Haruo. 1970. *Nez Perce Grammar*. University of California Publications, Linguistics 62. University of California Press: Berkeley.

Armstrong, Lilias E. 1964. *The Phonetic Structure of Somali*. Gregg Press: England.

Aronson, Howard I. 1989. *Georgian: A Reading Grammar*. Corrected Edition. Slavica Publishers, Inc: Columbus, Ohio.

Bahner, Werner, Joachim Schildt and Dieter Viehweger (Eds.). 1987. *Proceedings of the XIVth International Congress of Linguists II*. Akademie-Verlag: Berlin.

Baldi, Philip. 1983. *An Introduction to the Indo-European Languages*. Southern Illinois University Press: Carbondale and Edwardsville.

Bańczerowski, Jerzy. 1978. *A Formal Approach to the Theory of Fortition-Lenition: a Preliminary Study*. In Fisiak, Jacek (Ed.). "Recent Developments in Historical Phonology". 57-98.

Bandhu, C. M. and B. M. Dahal. 1971. *Nepali Segmental Phonology*. Summer Institute of Linguistics. Tribhuvan University: Kirtipur.

Barashkov, P. P. 1953. *Zvukovoj Sostav Jakutskovo Jazyka* (Sound System of the Yakut Language). Yakutsk.

Barnes, Michael. 2001. *Faroese Language Studies*. Studia Nordica 5. Novus Forlag: Oslo, Norway.

Bauer, Winifred. 1993. *Maori*. Routledge: London and New York.

Bayraktaroğlu, Arin and Sinan Bayraktaroğlu. 1992. *Colloquial Turkish*. Routledge: London and New York.

Beg, Mirza Khalil Ahmad. 1988. *Urdu Grammar: History and Structure*. Series in Indian Languages and Linguistics 18. Bahri Publications: New Delhi.

Bell, Alan, Daniel Jurafsky, Eric Fosler-Lussier, Cynthia Girand, Michelle Gregory, and Daniel Gildea. 2003. "Effects of Disfluencies, Predictability, and Utterance Position on Word Form Variation in English Conversation". *The Journal of the Acoustical Society of America*. 113:2: 1001-1024.

Bennett, William Holmes. 1980. *An Introduction to the Gothic Language*. The Modern Language Association of America: New York, NY.

Berman, Howard. 1990. "An Outline of Kalapuya Historical Phonology". *International Journal of American Linguistics* 56:1: 27-59. The University of Chicago Press: Chicago.

Best, Günter. 1983. *Culture and Language of the Turkana, NW Kenya*. Carl Winter: Heidelberg.

Bills, Garland D., Bernardo Vallejo C. and Rudolph C. Troike. 1969. *An Introduction to Spoken Bolivian Quechua*. Institute of Latin American Studies, The University of Texas Press: Austin and London.

Blevins, Juliette. 1995. *The Syllable in Phonological Theory*. In Goldsmith, John A. (Ed.). "The Handbook of Phonological Theory". 206-244.

Bliese, Loren F. 1981. *A Generative Grammar of Afar*. Poulter Virgil L. (Ed.). Summer Institute of Linguistics and The University of Texas at Arlington.

Bloch, Bernard. 1941. "Phonemic Overlapping". *American Speech* 16: 278-284. Reprinted in Joos, Martin (Ed.). 1957. "Readings in Linguistics: The Development of Descriptive Linguistics in America since 1925". 93-96.

Bokhari, Sohail. 1985. *Phonology of Urdu language*. Royal Book Company: Sadar Karachi.

Boltz, William G. and Michael C. Shapiro (Eds.). 1991. *Studies in the Historical Phonology of Asian Languages*. Volume 77. John Benjamins Publishing Company: Amsterdam and Philadelphia.

Bonvillain, Nancy. 1973. *A Grammar of Akwesasne Mohawk*. Mercury Series, Ethnology Division Paper No. 8. National Museum of Man, National Museums of Canada: Ottawa.

Borgman, Donald M. 1990. *Sanuma*. In Derbyshire, Desmond C. and Geoffrey K. Pullum (Eds.). "Handbook of Amazonian Languages, Volume 2". 15-248.

Brainard, Sherri. 1994. *The Phonology of Karao, The Philippines*. Pacific Linguistics Series B-110. Department of Linguistics, Research School of Pacific and Asian Studies, The Australian National University.

Brend, Ruth M. (Ed.). 1985. *From Phonology to Discourse: Studies in Six Colombian Languages*. Summer Institute of Linguistics: Dallas, Texas.

Brown, Gillian. 1972. *Phonological Rules and Dialect Variation: A Study of the Phonology of Lumasaaba*. Cambridge University Press: Great Britain.

Bubeník, Vít. 1983. *The Phonological Interpretation of Ancient Greek: A Pandialectal Analysis*. University of Toronto Press: Toronto, Canada.

Buck, Carl Darling. 1904. *A Grammar of Oscan and Umbrian*. Ginn and Company: Boston. (1970 authorized facsimile, University Microfilms, Ann Arbor, MI).

Burquest, Donald A. and Wyn D. Laidig (Eds.). 1992. *Phonological Studies in Four Languages of Maluku*. Publication 108. Summer Institute of Linguistics, The University of Texas at Arlington and Pattimura University.

Burrow, T. and S. Bhattacharya. 1970. *The Pengo Language: Grammar, Texts, and Vocabulary*. Oxford University Press: Oxford, Great Britain.

Bybee, Joan and Paul Hopper (Eds.). 2001. *Frequency and the Emergence of Linguistic Structure*. Typological Studies in Language 45. John Benjamins Publishing Company: Amsterdam and Philadelphia.

Bybee, Joan L. 1994. A View of Phonology from a Cognitive and Functional Perspective. *Cognitive Linguistics* 5-4: 285-305.

Callender, John B. 1975. *Middle Egyptian*. Afroasiatic Dialects Volume 2. Undena Publications: Malibu, California.

Campbell, Lyle. 1985. *The Pipil Language of El Salvador*. Mouton Publishers: Berlin, New York, Amsterdam.

Cardona, George. 1965. *A Gujarati Reference Grammar*. University of Pennsylvania Press: Philadelphia.

Cataloger's Reference Shelf, The. 1997-2001. Based on 21 MARC manuals and other reference works published by The Library of Congress. The Library Corporation: <http://www.tlcdelivers.com/tlc/crs/>.

Chisum, Gary L. 1975. *Systematic Phonology of the Kannada Noun.* In Schiffman, Harold F. and Carol M. Eastman (Eds.). "Dravidian Phonological Systems". 197-205.

Chomsky, Noam and Morris Halle. 1968. *The Sound Pattern of English.* Harper & Row Publishers: New York, Evanston, and London.

Chomsky, Noam. 1964. *Current Issues in Linguistic Theory.* In Fodor, Jerry A. and Jerrold J. Katz. "The Structure of Language: Readings in the Philosophy of Language". 50-118.

Christmas, R. B. and J. E. 1975. *Kupia Phonemic Summary.* Summer Institute of Linguistics, Institute of Nepal and Asian Studies, Tribhuvan University. University Press: Tribhuvan University, Kathmandu.

Claesson, Kenneth. 1994. "A Phonological Outline of Mataco-Noctenes". *International Journal of American Linguistics* 60:1: 1-38. The University of Chicago Press: Chicago.

Coetzee, Andries W. 1999. *Tiberian Hebrew Phonology: Focussing on Consonant Clusters.* Studiea Semitica Neerlandica. Van Gorcum: Assen, The Netherlands.

Colarusso, John. 1988. *The Northwest Caucasian Languages: A Phonological Survey.* Garland Publishing: New York and London.

Cole, Jennifer S. and José I. Hualde. 1998. "The Object of Lexical Acquisition: A UR-free Model". *Proceedings of Chicago Linguistic Society (CLS)* 34.

Collins, Beverly and Inger M. Mees. *The Phonetics of Cardiff English.* In Coupland, Nikolas (Ed.). 1990. "English in Wales: Diversity, Conflict and Change". 87-103.

Cook, Eung-Do and Jonathan Kaye (Eds.). 1978. *Linguistic Studies of Native Canada.* University of British Columbia Press: Vancouver, BC.

Costa, David J. 1991. "The Historical Phonology of Miami-Illinois Consonants". *International Journal of American Linguistics* 57:3: 365-393. The University of Chicago Press: Chicago.

Counts, David R. 1969. *A Grammar of Kaliai-Kove.* University of Hawaii Press.

Coupland, Nikolas (Ed.). 1990. *English in Wales: Diversity, Conflict and Change.* Multilingual Matters: Clevedon.

Cravens, Thomas D. 1983. *The Phonological Unity of Intervocalic Stop Weakening in Romance.* Doctoral Dissertation: University of Illinois at Urbana-Champaign.

Crowley, Terry. 1983. *Uradhi.* In Dixon R. M. W. and Barry J. Blake (Eds.). "The Handbook of Australian Languages Volume 3". 307-428.

Das, A. R. 1977. *A Study on the Nicobarese Language.* Anthropological Survey of India, government of India: Calcutta.

Dayley, Jon P. 1989a. *Tümpisa (Panamint) Shoshone Dictionary.* University of California Publications in Linguistics, Volume 116. University of California Press: Berkeley.

Dayley, Jon P. 1989b. *Tümpisa (Panamint) Shoshone Grammar.* University of California Publications in Linguistics, Volume 115. University of California Press: Berkeley.

de Wolf, Gaelan Dodds and Erika Hasebe-Ludt. 1987. *The Relationship of Speakers' Language Attitudes to Intervocalic Voicing in Formal and Informal Speech.* In Bahner, Werner, Joachim Schildt and Dieter Viehweger, (Eds.). "Proceedings of the XIVth International Congress of Linguists II". 1562-1567.

Dench, Alan. 1991. *Panyjima.* In Dixon R. M. W. and Barry J. Blake (Eds.). "The Handbook of Australian Languages Volume 4". 125-243.

Derbyshire, Desmond C. and Geoffrey K. Pullum (Eds.). 1986. *Handbook of Amazonian Languages, Volume 1.* Mouton de Gruyter: Berlin, New York, Amsterdam.

Derbyshire, Desmond C. and Geoffrey K. Pullum (Eds.). 1990. *Handbook of Amazonian Languages, Volume 2.* Mouton de Gruyter: Berlin and New York.

Derbyshire, Desmond C. and Geoffrey K. Pullum (Eds.). 1991. *Handbook of Amazonian Languages, Volume 3.* Mouton de Gruyter: Berlin and New York.

Diffloth, Gérard F. 1975. *The South Dravidian Obstruent System in Irula.* In Schiffman, Harold F. and Carol M. Eastman (Eds.). "Dravidian Phonological Systems". 47-56.

Dimmendaal, Gerrit Jan. 1983. *The Turkana Language.* Foris Publications: Dordrecht, Holland.

Dinnsen, Daniel A. (Ed.). 1979. *Current Approaches to Phonological Theory.* Indiana University Press: Bloomington & London.

Dinnsen, Daniel A. and Fred R. Eckman. 1977. "Some Substantive Universals in Atomic Phonology". *Lingua* 45: 1-14. North-Holland Publishing Company: Amsterdam, The Netherlands.

Dixon R. M. W. and Barry J. Blake (Eds.). 1983. *The Handbook of Australian Languages Volume 3.* The Australian National University Press: Canberra.

Dixon R. M. W. and Barry J. Blake (Eds.). 1991. *The Handbook of Australian Languages Volume 4.* Oxford University Press: Australia.

Dixon R. M. W. and Barry J. Blake (Eds.). 2000. *The Handbook of Australian Languages Volume 5.* Oxford University Press: Melbourne, Australia.

Dixon, R. M. W. 1983. *Nyawaygi.* In Dixon R. M. W. and Barry J. Blake (Eds.). "The Handbook of Australian Languages Volume 3". 431-525.

Dixon, R. M. W. 1991. *Mbabaram.* In Dixon R. M. W. and Barry J. Blake (Eds.). "The Handbook of Australian Languages Volume 4". 349-402.

Doke, Clement Martyn and S. M. Mofokeng. 1957. *Textbook of Southern Sotho Grammar.* Longmans, Green and Co.: London.

Driem, George van. 1987. *A Grammar of Limbu*. Mouton de Gruyter: Berlin, New York, Amsterdam.

Dunn, Leone. 1988. "Badimaya, A Western Australian Language". In *Papers in Australian Linguistics No. 17*. Pacific Linguistics Series A. 71: 19-149. The Australian National University: Canberra, Australia.

Dunstan, Elizabeth (Ed.). 1969. *Twelve Nigerian Languages*. Africana publishing corporation: New York.

Elbert, Samuel H. and Mary Kawena Pukui. 1979. *Hawaiian Grammar*. The University Press of Hawaii: Honolulu.

Elkins, Richard E. 1970. *Major Grammatical Patterns of Western Bukidnon Manobo*. Publication 26. Summer Institute of Linguistics of the University of Oklahoma.

Elson, Benjamin F. (Ed.). 1986. *Language in Global Perspective: Papers in Honor of the 50th Anniversary of the Summer Institute of Linguistics 1935-1985*. Summer Institute of Linguistics: Dallas, Texas.

Emeneau, M. B. 1951. *Studies in Vietnamese (Annamese) Grammar*. University of California Publications in Linguistics Volume 8. University of California Press: Berkeley and Los Angeles.

Encyclopedia of the Orient. <http://lexicorient.com/e.o/index.htm>.

Fisiak, Jacek (Ed.). 1978. *Recent Developments in Historical Phonology*. Mouton Publishers: The Hague.

Flemming, Edward. 1995. *Auditory Features in Phonology*. Doctoral Dissertation: University of California, Los Angeles.

Fodor, Jerry A. and Jerrold J. Katz. 1964. *The Structure of Language: Readings in the Philosophy of Language*. Prentice Hall, Inc.: Englewood Cliffs, New Jersey.

Fortescue, Michael. 1984. *West Greenlandic*. Croom Helm Descriptive Grammars. Croom Helm: Dover, New Hampshire.

Frantz, Donald G. 1971. *Toward a Generative Grammar of Blackfoot (with Particular Attention to Selected Stem Formation Processes)*. Davis Ervine (Ed.). Summer Institute of Linguistics of the University of Oklahoma.

Furbee-Losee, Louanna. 1976. *The Correct Language: Tojolabal. A Grammar with Ethnographic Notes*. Garland Publishing: New York & London.

Gerdel, Florence L. 1985. *Páez: Pitch and Stress in the Phonological Word and Phrase*. In Brend, Ruth M. (Ed.). "From Phonology to Discourse: Studies in Six Colombian Languages". 31-42.

Gill, Harjeet Singh and Henry A. Gleason, Jr. 1969. *A Reference Grammar of Punjabi*. Punjabi University Press: Patiala.

Gleason, Jean Berko (Ed.). 1993. *The Development of Language*. Macmillan Publishing Company: New York, NY.

Goldsmith, John A. (Ed.). 1995. *The Handbook of Phonological Theory*. Blackwell Publishers: Cambridge, MA.

Goswami, Golok Chandra. 1966. *An Introduction to Assamese Phonology*. Deccan College: Poona, India.

Greenberg, Joseph H. (Ed.). 1963. *Universals of Language*. Second Edition. MIT Press: Cambridge MA.

Greenberg, Joseph H. 1948. "The Tonal System of Proto-Bantu". *Word* 4:3: 196-208.

Greenberg, Joseph H. 1963. *Some Universals of Grammar with Particular Reference to the Order of Meaningful Elements*. In Greenberg, Joseph H. (Ed.). "Universals of Language". 73-113.

Greenberg, Joseph H. 1966. *Language Universals, With Special reference to Feature Hierarchies*. Mouton & Company: The Hague, The Netherlands.

Gregory, Michelle L. 2001. *Linguistic Informativeness and Speech Production: An Investigation of Contextual and Discourse-pragmatic Effects on Phonological Variation*. Doctoral Dissertation: University of Colorado at Boulder.

Grimes, Barbara F. (Ed.). 2002. *Ethnologue: Languages of the World*. 14th Edition. SIL International. <http://www.ethnologue.com/>.

Gudschinsky, Sarah C. and Harold and Frances Popovich. 1970. "Native Reaction and Phonetic Similarity in Maxakalí Phonology". *Language* 46:1: 77-88.

Guitart, Jorge M. 1976. *Markedness and a Cuban Dialect of Spanish*. Georgetown University Press: Washington DC.

Gurevich, Naomi. 2001. "A Critique of Markedness-Based Theories in Phonology". *Studies in the Linguistic Sciences* 31:2: 89-114.

Gurubasave Gowda, K. S. 1975. *Ao Grammar*. Central Institute of Indian Languages Grammar Series-1. E. Annamalai (Ed.). D. P. Pattanayak, Mysore-6. Kapila Power Press: Nanjangud, Karnataka State, India.

Hahn, Reinhard F. 1991. *Spoken Uyghur*. University of Washington Press: Seattle and London.

Halle, Morris and G. N. Clements. 1983. *Problem Book in Phonology*. MIT Press: Cambridge, MA and London, England.

Halle, Morris. 1964a. *On the Bases of Phonology*. In Fodor, Jerry A. and Jerrold J. Katz. "The Structure of Language: Readings in the Philosophy of Language". 324-333.

Halle, Morris. 1964b. *Phonology in Generative Grammar*. In Fodor, Jerry A. and Jerrold J. Katz. "The Structure of Language: Readings in the Philosophy of Language". 334-352.

Handbook of Western Australian Aboriginal Languages. <http://coombs.anu.edu.au/WWWVLPages/AborigPages/LANG/WA/contents.htm>.

Hardwick, Margaret F. 1984. *Tahltan Phonology and Morphology*. Ms.

Hargus, Sharon Louise. 1985. *The Lexical Phonology of Sekani.* Doctoral Dissertation: University of California, Los Angeles. University Microfilms International: Ann Arbor, MI.

Harms, Robert Thomas. 1962. *Estonian Grammar.* Indiana University Publications, Uralic and Altaic Series, Vol. 12. Indiana University: Bloomington.

Harrell, Richard S. 1957. *The Phonology of Colloquial Egyptian Arabic.* American Council of Learned Societies: New York.

Harris, James W. 1969. *Spanish Phonology.* Research Monograph No. 54. The MIT Press: Cambridge, MA and London, England.

Harris, John. 1990. "Segmental Complexity and Phonological Government". *Phonology* 7: 255-300. Cambridge University Press: Great Britain.

Harris, Martin and Nigel Vincent. 1987. *The Romance Languages.* Oxford University Press: New York.

Haspelmath, Martin. 1993. *A Grammar of Lezgian.* Mouton Grammar Library 9. Mouton de Gruyter: Berlin, New York.

Hayes, Bruce. 1986a. "Inalterability in CV Phonology". *Language* 62:2: 321-351. Linguistic Society of America: Baltimore, MD.

Hayes, Bruce. 1986b. "Assimilation as Spreading in Toba Batak". *Linguistic Inquiry* 17:3: 467-499.

Hayes, Bruce. 1999. *Phonetically Driven Phonology: The Role of Optimality Theory and Inductive Grounding.* In Darnell, Michael, Edith Moravscik, Michael Noonan, Frederick Newmeyer, and Kathleen Wheatly (Eds.). "Functionalism and Formalism in Linguistics, Volume I: General Papers". John Benjamins: Amsterdam. 243-285.

Hayward, Dick. 1984. *The Arbore Language: A First Investigation.* Helmut Buske Verlag: Hamburg.

Heath, Jeffrey. 1980. *Basic Materials in Warndarang: Grammar, Texts and Dictionary.* Pacific Linguistics Series B, No. 72. Department of Linguistics, Research School of Pacific Studies, The Australian National University: Canberra, Australia.

Hebert, Raymond J. and Nicholas Poppe. 1963. *Kirghiz Manual.* Indiana University Publications. Uralic and Altic Series, Vol. 33. Indiana University: Bloomington.

Herbert, Robert K. 1986. "Language Universals, Markedness Theory, and Natural Phonetic Process". *Trends in Linguistics, Studies and Monographs 25.* Winter Werner (Ed.). Mouton de Gruyter.

Hoard, James E. 1978. *Obstruent Voicing in Gitksan: Some Implications for Distinctive Feature Theory.* In Cook Eung-Do and Jonathan Kaye (Eds.). "Linguistic Studies of Native Canada". 111-120.

Hock, Hans Henrich and Brian D. Joseph. 1996. *Language History, Language Change and Language Relationship. an Introduction to Historical and Comparative Linguistics*. Mouton de Gruyter: Berlin and New York.

Hock, Hans Henrich. 1976. *Final Weakening and Related Phenomena*. In Ingemann, Frances (Ed.). "1975 Mid-America Linguistics Conference Papers". The Linguistics Department, University of Kansas: Lawrence, Kansas.

Hock, Hans Henrich. 1985. "Regular Metathesis". *Linguistics* 23:4: 529-546.

Hock, Hans Henrich. 1991a. *Dialects, Diglossia, and Diachronic Phonology in Early Indo-Aryan*. In Boltz, William G. and Michael C. Shapiro (Eds.). "Studies in the Historical Phonology of Asian Languages". 119-159.

Hock, Hans Henrich. 1991b. *Principles of Historical Linguistics*. 2nd Ed., Revised and Updated. Mouton de Gruyter: Berlin and New York.

Hock, Hans Henrich. 1992. *Initial Strengthening*. In Dressler, Wolfgang U., Hans C. Luschützky, Oskar E. Pfeiffer and John R. Rennison (Eds.). "Phonologica 1988: Proceedings of the 6th International Phonology Meeting". Cambridge University Press: Great Britain. 101-110.

Hock, Hans Henrich. 1998. *In the End is my Beginning: Finality, Prosody, and Change*. MS., University of Illinois at Urbana-Champaign.

Hoehlig, Monika and Maria Hari. 1976. *Kagate Phonemic Summary*. Summer Institute of Linguistics. Institute of Nepal and Asian Studies, Tribhuvan University: Kathmandu, Nepal.

Hoenigswald, Henry M. 1946. "Sound Change and Linguistic Structure". *Language* 22: 138-143.

Holmes, Janet. 1994. "New Zealand Flappers: an Analysis of T Voicing in New Zealand English". *English World-Wide* 15:2: 195-224. John Benjamins: Amsterdam.

Houlihan, Kathleen and Gregory K. Iverson. 1979. *Functionally-Constrained Phonology*. In Dinnsen, D. A. (Ed.). "Current Approaches to Phonological Theory". 50-73.

Hsu, Chai-Shune Katrina. 1995. *A Phonetically-Based Optimality Theoretic Account of Consonant Reduction in Taiwanese*. Masters of Arts Thesis: University of California, Los Angeles.

Hualde, José Ignacio and Jon Ortiz de Urbina (Eds.). 1993. *Generative Studies in Basque Linguistics*. Current Issues in Linguistic Theory, 105. John Benjamins Publishing Company: Amsterdam and Philadelphia.

Hualde, José Ignacio, Gorka Elordieta and Arantzazu Elordieta. 1994. *The Basque Dialect of Lekeitio*. Supplements of Anuario del Seminario de Filología Vasca «Julio de Urquijo», XXXIV. University of the Basque Country.

Hualde, José Ignacio. 1992. *Catalan*. Routledge: London and New York.

Hualde, José Ignacio. 1993. *Topics in Souletin Phonology*. In Hualde, José Ignacio and Jon Ortiz de Urbina (Eds.). "Generative Studies in Basque Linguistics". 289-327.

Hualde, José Ignacio. 2000. "On System-Driven Sound Change: Accent Shift in Markina Basque". *Lingua* 110: 99-129.

Hudak, Thomas John (Ed.). 1993. *William J. Gedney's The Saek Language: Glossary, Texts, and Translations*. Michigan Papers on South and Southeast Asia, Number 41. Center for South and Southeast Asian Studies, The University of Michigan: Ann Arbor, Michigan.

Hyman, Larry M. 1999. *Contexts of Fortition and Lenition in Bantu*. Abstract. Presented at the "International Conference in Phonology" in Nice, France.

Ingemann, Frances (Ed.). 1976. *1975 Mid-America Linguistics Conference Papers*. The Linguistics Department, University of Kansas: Lawrence, Kansas.

Ito, Junko. 1989. "A Prosodic Theory of Epenthesis". *Natural Language and Linguistic Theory* 7: 217-259.

Jacobs, Haike and Leo Wetzels. 1988. *Early French Lenition: a Formal Account of an Integrated Sound Change*. In van der Hulst, Harry and Norval Smith (Eds.). "Features, Segmental Structure and Harmony Processes". Part I. 105-129.

Jacobs, Haike. 1994. *Lenition and Optimality Theory*. Proceedings of LSRL XXIV, Feb. 1994. Rutgers Optimality Archive. ROA# 127-0496. <http://roa.rutgers.edu/>.

Jacobs, Haike. 1995. Optimality Theory and Sound Change. Paper presented at NELS 25, 1995. Rutgers Optimality Archive. ROA# 129-0496. <http://roa.rutgers.edu/>.

Jespersen, Otto. 1964. *Language: Its Nature, Development and Origin*. W.W. Norton & Company: New York.

Joos, Martin (Ed.). 1957. *Readings in Linguistics: The Development of Descriptive Linguistics in America since 1925*. American Council of Learned Societies: Washington.

Joos, Martin. 1942. "A Phonological Dilemma in Canadian English". *Language* 18: 141-144.

Jun, Jongho. 1995. *Perceptual and Articulatory Factors in Place Assimilation: an Optimality Theoretic Approach*. Doctoral Dissertation: University of California, Los Angeles.

Jurafsky, Daniel, Alan Bell, Michelle Gregory and William D. Raymond. 2001. *Probabilistic Relations between Words: Evidence from Reduction in Lexical Production*. In Bybee, Joan and Paul Hopper (Eds.). "Frequency and the Emergence of Linguistic Structure". 229-254.

Kachru, Braj B. 1969. *A Reference Grammar of Kashmiri*. Department of Linguistics, University of Illinois at Urbana-Champaign.

Kahn, Daniel. 1976. *Syllable-based Generalizations in English Phonology*. Doctoral Dissertation: Massachusetts Institute of Technology. Indiana University Linguistics Club: Bloomington, Indiana.

Kakumasu, James. 1986. *Urubu-Kaapor*. In Derbyshire, Desmond C. and Geoffrey K. Pullum (Eds.). "Handbook of Amazonian Languages, Volume 1". 326-403.

Kari, James M. 1976. *Navajo Verb Prefix Phonology*. Garland Publishing, Inc.: New York and London.

Kaufman, Terrence. 1971. *Tzeltal Phonology and Morphology*. University of California Publications, Linguistics 61. University of California Press: Berkeley.

Kaye, Jonathan Derek. 1979. *On the Alleged Correlation of Markedness and Rule Function*. In Dinnsen, D. A. (Ed.). "Current Approaches to Phonological Theory". 272-280.

Keating, Patricia, Wendy Linker and Marie Huffman. 1983. "Patterns in Allophone Distribution for Voiced and Voiceless Stops". *Journal of Phonetics* 11: 277-290. Academic Press: London.

Keels, Jack. 1985. *Guayabero: Phonology and Morphophonemics*. In Brend, Ruth M (Ed.). "From Phonology to Discourse: Studies in Six Colombian Languages". 57-87.

Kenstowicz, Michael and Charles Kisseberth. 1979. *Generative Phonology*. Academic Press, Inc.: San Diego, CA.

Kenstowicz, Michael. 1982. "Gemination and Spirantization in Tigrinya". *Studies in the Linguistic Sciences* 12:1: 103-122.

Kenstowicz, Michael. 1994. *Phonology in Generative Grammar*. Blackwell: Cambridge, MA.

Kenstowicz, Michael. 2000. *Paradigmatic Uniformity and Contrast*. <http://web.mit.edu/linguistics/www/kenstowicz/paradigm_contrast.doc>.

Khalafallah, Abdelghany A. 1969. *A Descriptive Grammar of Saʕi:di Egyptian Colloquial Arabic*. Mouton: The Hague and Paris.

Kiparsky, Paul. 1968. *How Abstract is Phonology?* Reprinted in "Explanation in Phonology". 1982. Foris Publications: The Netherlands. 119-163.

Kiparsky, Paul. 1995. *The Phonological Basis of Sound Change*. In Goldsmith, John A. (Ed.). "The Handbook of Phonological Theory". 640-670.

Kirchner, Robert. 1995. *Contrastiveness Is an Epiphenomenon of Constraint Ranking*. Proceedings of the Berkeley Linguistics Society 21.

Kirchner, Robert. 1998. *An Effort-Based Approach to Consonant Lenition*. Doctoral Dissertation: University of California, Los Angeles.

Kirchner, Robert. 2000. *Geminate Inalterability and Lenition*. Rutgers Optimality Archive. ROA# 249-0398. <http://roa.rutgers.edu/>.

Koehn, Edward and Sally Koehn. 1986. *Apalai*. In Derbyshire, Desmond C. and Geoffrey K. Pullum (Eds.). "Handbook of Amazonian Languages, Volume 1". 33-127.

Koshal, Sanyukta. 1976. *Ladakhi Phonetic Reader*. Central Institute of Indian Languages Phonetic Reader Series 18. B. G. Misra (Ed.). Manasagangotri, Mysore-570004, Karnataka State, India.

Krueger, John R. 1962. *Yakut Manual*. Indiana University Publications, Uralic and Altaic Series, Volume 21. Indiana University: Bloomington and Mouton & Co.: The Hague, The Netherlands.

Krysstal Web Site, The. 1997-2002. <http://www.krysstal.com/>. Kryss Katsiavriades and Talaat Qureshi in Association with Amazon and Britannica.

Labov, William. 1994. *Principles of Linguistic Change*. Blackwell: Oxford, UK and Cambridge, MA.

Lavoie, Lisa M. 1996. *Consonant Strength: Results of a Data Base Development Project*. Working Papers of the Cornell Phonetics Laboratory Vol. 11: 269-316.

Leal, Dorothy. 1972. *Chitwan Tharu Phonemic Summary*. Nepal Indo-Aryan Series I. Summer Institute of Linguistics and Institute of Nepal Studies: Tribhuvan University, Kirtipur, Nepal.

Lieberman, Philip. 1963. "Some Effects of Semantic and Grammatical Context on the Production and Perception of Speech". *Language and Speech* 6:3: 172-187.

Lin, Yen-Hwei. 1989. *Autosegmental Treatment of Segmental Processes in Chinese Phonology*. Doctoral Dissertation: University of Texas at Austin.

Lipski, John M. 1984. "On the Weakening of /s/ in Latin American Spanish". *Zeitschrift für Dialektologie und Linguistik* 51:1: 31-43.

Lockwood, W. B. 1964. *An Introduction to Modern Faroese*. Munksgaard: Copenhagen, Denmark.

Lowenstamm, Jean and Jonathan Kaye. 1986. *Compensatory Lengthening in Tiberian Hebrew*. In Wetzels, Leo and Engin Sezer (Eds.). "Studies in Compensatory Lengthening". 97-132.

Lukas, Johannes. 1967. *A Study of the Kanuri Language*. Dawsons of Pall Mall: London.

MacDonald, Lorna. 1990. *A Grammar of Tauya*. Mouton de Gruyter: Berlin and New York.

MacKay, Carolyn J. 1994. "A Sketch of Misantla Totonac Phonology". *International Journal of American Linguistics* 60:4: 369-419. The University of Chicago Press: Chicago.

Malone, Joseph L. 1993. *Tiberian Hebrew Phonology*. Eisenbrauns: Winona Lake, Indiana.

Manaster Ramer, Alexis. 1984. "Kern laws". *International Journal of American Linguistics* 50: 154-60. The University of Chicago Press: Chicago.

Manaster Ramer, Alexis. 1993. "On Lenition in Some Northern Uto-Aztecan Languages". *International Journal of American Linguistics* 59:3: 334-341. The University of Chicago Press: Chicago.

Manessy, Gabriel. 1969. *Les Langues Gurunsi*. Paris: SELAF.

Marshall, Margaret, M. 1984. *The Dialect of Notre-Dame-De-Sanilhac: A Natural Generative Phonology*. Stanford French and Italian Studies 31. Anma Libri: Saratoga, CA.

Martin, Samuel E. 1992. *A Reference Grammar of Korean: A Complete Guide to the Grammar and History of the Korean Language*. Charles E. Tuttle Company: Rutland, Vermont & Tokyo, Japan.

Martinet, A. 1952. "Function, Structure, and Sound Change". *Word* 8:2: 1-32.

Martinet, A. 1962. *A Functional View of Language*. Clarendon Press: Oxford.

McGregor, William. 1990. *A Functional Grammar of Gooniyandi*. John Benjamins Publishing Company: Amsterdam.

McLaughlin, John E. 1989. "A Note on the Change of Strident to Nonstrident in Gosiute Shoshoni". *International Journal of American Linguistics* 55:2: 240-247. The University of Chicago Press: Chicago.

McMahon, April M. S. 1994. *Understanding Language Change*. Cambridge University Press: Great Britain.

Meillet, Antoine. 1970. *General Characteristics of the Germanic Languages*. Translated by William P. Dismukes. Miami Linguistics Series No. 6. University of Miami Press: Coral Gables, Florida.

Mills, Elizabeth. 1984. *Senoufo Phonology, Discourse to Syllable (a Prosodic Approach)*. Publication Number 72. Summer Institute of Linguistics and University of Texas at Arlington.

Milroy James, Lesley Milroy and Sue Hartley. 1994. "Local and Supra-local Change in British English: The Case of globalization". *English World-Wide* 15:1: 1-33. John Benjamins B.V.: Amsterdam.

Mitleb, F. M. 1981. *Segmental and Non-segmental Structure in Phonetics: Evidence from Foreign Accent*. Doctoral Dissertation: Indiana University, Bloomington.

Mohanan, Karuvannur P. 1986. *The Theory of Lexical Phonology*. Studies in Natural Language and Linguistic Theory. D. Reidel Publishing Company, Kluwer Academic Publishers Group: Dordrecht, Holland.

Morev, L. N., A. A. Moskalyov, and Y. Y. Plam. 1979. *The Lao Language*. Nauka Publishing House, Central Department of Oriental Literature: Moscow.

Morey, Virginia and Ursula Post. 1964. *The phonology of Ata*. Unpublished paper in SIL Philippines files (cited in Reid 1971).

Nanda, Ishwara. (Strestahacharya, Iswaranand, Jagan Nath Maskey, and Austin Hale). 1971. *Conversational Newari.* Summer Institute of Linguistics, Institute for Nepal Studies, Tribhuvan University: Kathmandu, Nepal.

Nater, Hank F. 1989. "Some Comments on the Phonology of Tahltan". *International Journal of American Linguistics* 55:1: 25-42. The University of Chicago Press: Chicago.

Newman, Paul. 1974. *The Kanakuru Language.* West African Language Monographs 9. Institute of Modern English Language Studies, University of Leeds, in association with the West African Linguistic Society. The Scolar Press Ltd: Ilkley, Yorkshire, Great Britain.

Nivens, Richard. 1992. *Lexical Phonology of West Tarangan.* In Burquest, Donald A. and Wyn D. Laidig (Eds.). "Phonological Studies in Four Languages of Maluku". 127-227.

Noss, Richard B. 1964. *Thai Reference Grammar.* Department of State, Foreign Service Institute: Washington, D.C.

Nuñez Cedeño, Rafael A. 1987. "Intervocalic /d/ Rhotacism in Dominican Spanish: A Non Linear Analysis". *Hispania* 70:2: 363-368.

Ó Siadhail, Mícheál. 1989. *Modern Irish: Grammatical Structure and Dialectal Variation.* Cambridge University Press: Cambridge.

O'Grady, Geoffrey N. 1966. "Proto-Ngayarda Phonology". *Oceanic Linguistics* 5:2: 71-130.

Oftedal, Magne. 1985. *Lenition in Celtic and in Insular Spanish: The Secondary Voicing of Stops in Gran Canaria.* Monographs in Celtic Studies from the University of Oslo, Volume 2. Universitetsforlaget: Oslo.

Ourso, Meterwa A. and Charles H. Ulrich. 1990. "Sonorant-strengthening in Lama". *Studies in the Linguistic Sciences* 20:1: 135-147.

Oxford English Dictionary, The. 2003. <http://dictionary.oed.com/>. Oxford University Press.

Oxford Reference Online. 2002. <http://www.oxfordreference.com/>. Oxford University Press.

Padgett, Jaye. 1999. *Contrast Dispersion and Russian Palatalization.* (to appear) in Hume Elizabeth and Keith Johnson (Eds.). "The Role of Perceptual Phenomena in Phonological Theory". Academic Press.

Paradis, Carole and Jean-François Prunet (Eds.). 1991. *The Special Status of Coronals: Internal and External Evidence.* Phonetics and Phonology, Volume 2. Academic Press, Inc.: San Diego, CA.

Parks, Douglas, R. 1976. *A Grammar of Pawnee.* Garland Publishing, Inc.: New York & London.

Partridge, Monica. 1988. *Serbo-Croat Practical Grammar and Reader.* Prosveta: Beograd.

Patz, Elisabeth. 1991. *Djabugay.* In Dixon R. M. W. and Barry J. Blake (Eds.). 1991. "The Handbook of Australian Languages Volume 4". 245-348.

Penny, Ralph John. 2002. *A History of the Spanish Language*. Cambridge University Press: Cambridge and New York.

Pike, Eunice V., Keith Forester and Wilma J. Forester. 1986. *Fortis Versus Lenis Consonants in the Paya Dialect of Kuna*. In Elson, Benjamin F. (Ed.). "Language in Global Perspective: Papers in Honor of the 50th Anniversary of the Summer Institute of Linguistics 1935-1985". 451-464.

Popjes, Jack and Jo Popjes. 1986. *Canela-Krahô*. In Derbyshire, Desmond C. and Geoffrey K. Pullum (Eds.). "Handbook of Amazonian Languages, Volume 1". 128-199.

Poppe, Nicholas. 1964. *Bashkir Manual*. Indiana University Publications, Uralic and Altaic Series, Vol. 36. Indiana University, Bloomington.

Poppe, Nicholas. 1968. *Tatar Manual: Descriptive Grammar and Texts with A Tatar-English Glossary*. Uralic and Altaic Series, Volume 25 (2nd, Revised Edition). Indiana University: Bloomington & Mouton & Co.: The Hague, The Netherlands.

Poppe, Nicholas. 1970. *Mongolian Language Handbook*. Center for Applied Linguistics: Washington, D.C.

Port, R. and P. Crawford. 1989. "Pragmatic Effects on Neutralization Rules". *Journal of Phonetics* 16:4: 257-282.

Port, Robert F. and Michael L. O'Dell. 1985. "Neutralization of Syllable-Final Voicing in German". *Journal of Phonetics* 13:4: 455-471. Academic Press: London.

Postal, Paul M. 1968. *Aspects of Phonological Theory*. Harper and Row: New York, Evanston, and London.

Prentice, D. J. 1971. *The Murut Languages of Sabah*. Pacific Linguistics Series C, No. 18. The Australian National University: Canberra, Australia.

Prince, Alan and Paul Smolensky. 1993 (ROA Version 2002). *Optimality Theory: Constraint Interaction in Generative Grammar*. Rutgers Optimality Archive. ROA# 537-0802. <http://roa.rutgers.edu/>.

Prince, Alan Sanford. 1975. *The Phonology and Morphology of Tiberian Hebrew*. Doctoral Dissertation: Massachusetts Institute of Technology.

Prokosch, E. 1933. *An Outline of German Historical Grammar*. Oxford University Press: New York.

Pulleyblank, Edwin. G. 1984. *Middle Chinese: A Study in Historical Phonology*. University of British Columbia Press: Vancouver.

Rajapurohit, B. B. 1983. *Shina Phonetic Reader*. Central Institute of Indian Languages Phonetic Reader Series-24. E. Annamalai (Ed.). Manasagangotri, Mysore-570006.

Ramsey, Robert S. 1991. *Proto-Korean and the Origin of Korean Accent*. In Boltz, William G. and Michael C. Shapiro (Eds.). "Studies in the Historical Phonology of Asian Languages". 215-238.

Rangan, K. 1975. *Balti Phonetic Reader.* Central Institute of Indian Languages Phonetic Reader Series-17. E. Annamalai (Ed.). D. P. Pattanayak, Mysore-570006.

Rangan, K. 1979. *Purki Grammar.* Central Institute of Indian Languages Grammar Series-5. E. Annamalai (Ed.). Manasagangotri, Mysore-570006.

Read, A. F. C., Central Asian Mission. 1934. *Balti Grammar.* James G. Forlong Fund, Vol. XV. The Royal Asiatic Society: London.

Reid, Lawrence A. (Ed.). 1971. *Philippine Minor Languages: Word Lists and Phonologies.* Oceanic Linguistics Special Publication No. 8. Summer Institute of Linguistics and University of Hawaii. University of Hawaii Press: Honolulu, Hawaii.

Reid, Lawrence A. 1993. *Another Look at the Language of the Tasaday.* Keynote address, The Third Annual Conference of the Southeast Asian Linguistic Society. Honolulu, Hawai'i.

Resnick, Melvyn C. 1975. *Phonological Variants and Dialect Identification in Latin American Spanish.* Mouton: The Hague, Netherlands.

Rice, Keren. 1983a. *Slave Grammar.* MS.

Rice, Keren. 1983b. *Athapaskan Fricatives and Markedness Theory.* MS.

Rice, Keren. 1999. "Featural Markedness in Phonology: Variation". Part 1. *International GLOT* 4:7: 3-6. Part 2. *International GLOT* 4:8: 3-7.

Roberts, John R. 1987. *Amele.* Croom Helm: London, New York, Sydney.

Romero Gallego, Joaquín. 1995. *Gestural Organization in Spanish. An Experimental Study of Spirantization and Aspiration.* Doctoral Dissertation: University of Connecticut.

Saib, J. 1976. *A Phonological Study of Tamazight Berber: Dialect of the Ayt Ndhir.* Doctoral Dissertation, University of California, Los Angeles. Xerox University Microfilms: Ann Arbor, Michigan.

Samarin, William J. 1966. *The Gbeya Language: Grammar, Texts and Vocabularies..* University of California Press: Berkeley and Los Angeles.

Sapir, Edward and Morris Swadesh. 1960. *Yana Dictionary.* Haas, Mary R. (Ed.). University of California Publications in Linguistics Volume 22. University of California Press: Berkeley and Los Angeles.

Scatton, Ernest A. 1984. *A Reference Grammar of Modern Bulgarian.* Slavica Publishers, Inc.: Columbus, Ohio.

Schiffman, Harold F. and Carol M. Eastman (Eds.). 1975. *Dravidian Phonological Systems.* South Asian Studies Program, Institute for Comparative and Foreign Area Studies, University of Washington: Seattle.

Ségéral, Philippe and Tobias Scheer. 1999. *The Coda Mirror.* MS. Université Paris 7 and Université de Nice, France.

Seidenadel, Carl Wilhelm. 1909. *The Language Spoken by the Bontoc Igorot.* The Open Court Publishing Company: Chicago.

Sharma, D. D. 1982. *Studies in Tibeto-Himalayan Linguistics: A Descriptive Analysis of Pattani (A Dialect of Lahaul).* Vishveshvaranand Vishva Bandhu Institute of Sanskrit and Indological Studies, Panjab University: Hoshiarpur.

Sharma, J. D. 1979. *Gojri Phonetic Reader.* Central Institute of Indian Languages Series-19. E. Annamalai (Ed.). D. P. Pattanayak, Mysore-570006.

Shimkin, D. B. 1949. "Shoshone, I: Linguistic Sketch and Text". *International Journal of American Linguistics* 15:3: 175-188. Waverly Press, Inc.: Baltimore, MD.

Shipley, William F. 1963. *Maidu Texts and Dictionary.* University of California Publications in Linguistics, Volume 33. University of California Press: Berkeley and Los Angeles.

Silverman, Daniel. 1997. "Tone Sandhi in Comaltepec Chinantec". *Language* 73:3: 473-492.

Silverman, Daniel. 2000a. *Alveolar Stops in American English, and the nature of Allophony.* In Tamanji, Pius N. and Kiyomi Kusumoto (Eds.). NELS 28. GLSA, University of Massachusetts: Amherst.

Silverman, Daniel. 2000b. "Hypotheses Concerning the Phonetic and Functional Origins of Tone Displacement in Zulu". *Studies in African Linguistics* 29.2: 1-32. Indiana University: Bloomington.

Silverman, Daniel. 2002. "Dynamic Versus Static Phonotactic Constraints in Prosodic Morphology". *Linguistics* 40.1: 28-58.

Silverman, Daniel. 2003a (to appear). "On the Phonetic and cognitive Nature of Alveolar Stop Allophony in American English". *Cognitive Linguistics.*

Silverman, Daniel. 2003b (to appear). "On the Rarity of Pre-Aspirated Stops". *Journal of Linguistics.*

Singh, R. A. 1977. *Survey of Halabi in Madhya Pradesh.* Language Division, Office of the Registrar Central: India. Census of India 1971, Series 1, Monograph No. 5.

Sjoberg, Andrée F. 1963. *Uzbek Structural Grammar.* Uralic and Altaic Series, Vol. 18. Indiana University: Bloomington and Mouton & Co.: The Hague, The Netherlands.

Smeets, Rieks. 1984. *Studies in West Circassian Phonology and Morphology.* The Hakuchi Press: Leiden.

Smith, Neilson V. 1981. "Consistency, Markedness and Language Change: On the Notion 'Consistent Language'". *Journal of Linguistics* 17: 39-55.

Smith, Norval, Mieke Beers, Rens Bod, Roberto Bolognesi, Helga Humbert and Frank van der Leeuw. 1991. *Lenition in a Sardinian Dialect.* Certamen Phonologicum II: Papers from the 1990 Cortona Phonology Meeting. Rosenberg & Sellier: Torino. 309-328.

Sommerstein, Alan H. 1973. *The Sound pattern of Ancient Greek*. Publications of the Philological Society XXIII. Basil Blackwell: Oxford, England.

Stallcup, Kenneth L. 1978. *A comparative Perspective on the Phonology and Noun Classification of Three Cameroon Grassfields Bantu Languages: Moghamo, Ngie, and Oshie*. Doctoral Dissertation: Stanford University. University Microfilms International: Ann Arbor, MI.

Steriade, Donca. 1999. *Alternatives to the Syllabic Interpretation of Consonantal Phonotactics*. In Fujimura O., B. Joseph and B. Palek (Eds.). "Proceedings of the 1998 Linguistics and Phonetics Conference". The Karolinum Press. 205-242.

Steriade, Donca. 2000. *Paradigm Uniformity and the Phonetics/Phonology Boundary*. In Pierrehumbert, J. and M. Broe (Eds.). "Papers in Laboratory Phonology Vol. 6". Cambridge University Press. 313-335.

Story, Gillian L. 1984. *Babine & Carrier Phonology: A Historically Oriented Study*. Poulter Virgil L. and Desmond C. Derbyshire (Eds.). Summer Institute of Linguistics and The University of Texas at Arlington.

Strodach, Elizabeth Aborn. 1976. *Studies in Phonological and Analogical Change in Early North Germanic*. Doctoral Dissertation: University of California, Berkeley.

Sulkala, Helena and Merja Karjalainen. 1992. *Finnish*. Routledge: London.

Supple, Julia and Celia M. Douglass. 1949. "Tojolabal (Mayan): Phonemes and Verb Morphology". *International Journal of American Linguistics* 15:3: 168-174. Waverly Press, Inc.: Baltimore, MD.

Swift, L.B. 1963. *A Reference Grammar of Modern Turkish*. Indiana University: Bloomington.

Taylor, Charles. 1985. *Nkore-Kiga*. Croom Helm Descriptive Grammars. Croom Helm: London.

Teeter, Karl V. 1964. *The Wiyot Language*. University of California Publications in Linguistics Volume 37. University of California Press: Berkeley and Los Angeles.

Tekavčić, Pavao. 1972. *Grammatica Storica dell' Italiano. Vol. 1: Fonematica*. Bologna.

Teodorsson, Sven-Tage. 1977. *The Phonology of Ptolemaic Koine*. Studia Graeca et Latina Gothoburgensia XXXVI. Acta Universitatis Gothoburgensis: Göteborg, Sweden.

Thráinsson, Höskuldur. 1978. "On the Phonology of Icelandic Preaspiration". *Nordic Journal of Linguistics* 1: 3-54. Universitetsforlaget: Oslo, Norway.

Tinelli, Henri. 1981. *Creole Phonology*. Mouton: The Hague, Netherlands.

Tosco, Mauro. 1991. *A Grammatical Sketch of Dahalo*. Helmut Buske Verlag: Hamburg.

Trail, Ronald L. 1970. *The Grammar of Lamani.* A thesis submitted to the University of Poona for the degree of Doctor of Philosophy in Linguistics, 1968. Summer Institute of Linguistics, Publication Number 24. Summer Institute of Linguistics: Norman, Oklahoma.

Trubetzkoy, N.S. 1939. *Principles of Phonology.* Translated by Christiane A. M. Baltaxe. 1971. LA: University of California Press.

Tyler, Stephen A. 1975. *Gondi /h/ With Some Notes on Number and Gender in Proto-Central Dravidian.* In Schiffman, Harold F. and Carol M. Eastman (Eds.). "Dravidian Phonological Systems". 86-113.

UCLA Language Materials Project, The. Provided in collaboration with the Center for Applied Linguistics, Washington, DC (CAL Database). <http://www.lmp.ucla.edu/>.

Underhill, Robert. 1976. *Turkish Grammar.* The MIT Press: Cambridge, Massachusetts, and London, England.

van der Hulst, Harry and Norval Smith (Eds.). 1988. *Features, Segmental Structure and Harmony Processes.* Parts I and II. Foris Publications: Dordrecht.

Vennemann, Theo. 1972. *Sound Change and Markedness Theory: On the History of the German Consonant System.* In Stockwell, Robert P. and Ronald K. S. Macaulay (Eds.). "Linguistic Change and Generative Theory. Essays from the UCLA Conference on Historical Linguistics in the Perspective of Transformational Theory". Indiana University Press: Bloomington and London. 230-274.

Vennemann, Theo. 1988. *Preference Laws for Syllable Structure and the Explanation of Sound Change With Special Reference to German, Germanic, Italian, and Latin.* Mouton de Gruyter: Berlin.

Vinas Urquiza, Maria T. 1970. *Fonología de la Lengua Mataca.* Cuadernos de Lingüística Indígena, no. 7. Instituto de lingüística, Universidad de Buenos Aires: Buenos Aires.

Vincent, Nigel. 1978. *Is Sound Change Teleological?* In Fisiak, Jacek (Ed.). "Recent Developments in Historical Phonology". 409-430.

Vovin, Alexander. 1993. *A Reconstruction of Proto-Ainu.* E.J. Brill: Leiden, New York, Köln.

Wade, Terence. 1992. *A Comprehensive Russian Grammar.* Blackwell: Oxford, UK.

Wares, Alan Campbell. 1968. *A Comparative Study of Yuman Consonantism.* Summer Institute of Linguistics. Mouton: The Hague, Paris.

Weinrich, Harald. 1958. *Phonologische Studien zur Romanischen Sprachgeschichte = Forschungen zur romanischen Philologie.* Heft 6. Münster.

Wells, J. C. 1982a. *Accents of English 1: An Introduction*. Cambridge University Press: Cambridge.

Wells, J. C. 1982b. *Accents of English 2: The British Isles*. Cambridge University Press: Cambridge.

Wells, J. C. 1982c. *Accents of English 3: Beyond the British Isles*. Cambridge University Press: Cambridge.

Westbury, J. R. and P. A. Keating. 1986. "On the Naturalness of Stop Consonant Voicing". *Journal of Linguistics* 22: 145-166.

Wetzels, Leo and Engin Sezer (Eds.). 1986. *Studies in Compensatory Lengthening*. Foris Publications: Dordrecht.

Whisler, Ronald. 1992. *Phonology of Sawai*. In Burquest, Donald A. and Wyn D. Laidig (Eds.). "Phonological Studies in Four Languages of Maluku". 7-32.

Whitley, Stanley M. 1978. *Generative Phonology Workbook*. The University of Wisconsin Press: Madison, Wisconsin.

Wordick, F. J. F. 1982. *The Yindjibarndi Language*. Pacific Linguistics Series C No. 71. Department of Linguistics, Research School of Pacific Studies. The Australian National University: Canberra, Australia.

Wright, Richard Albert. 1996. *Consonant Clusters and Cue Preservation in Tsou*. Doctoral Dissertation: University of California, Los Angeles.

Yar-Shater, Ehsan. 1969. *A Grammar of Southern Tati Dialects*. Median Dialect Studies I. Mouton: The Hague.

Yip, Moira. 1991. *Coronals, Consonant Clusters, and the Coda Condition*. In Paradis, Carole and Jean-François Prunet (Eds.). "The Special Status of Coronals: Internal and External Evidence. Phonetics and Phonology, Volume 2". 61-78.

Zhang, Z. X. 1983. *Taiwan Minnan Fangyan Jilue* (Notes on Southern Min Dialects of Taiwan). Fujian People's Press: Fuzhou.

Zipf, G. K. 1935. *The Psycho-Biology of Language*. Houghton Mifflin: Boston.

APPENDIX A

Result Summary

Key: 1=Not Neutralization; 2=Limited Neutralization; 3=Incomplete Neutralization; 4=Neutralization;

Name	Phylum	Source	Process	Type	1	2	3	4
Amele (Haia) 1	Trans-New Guinea	Roberts 1987	p ~ f	spirantization	X			
Amele (Haia) 2			b ~ p	voicing	X			
Ancient Greek (Attic)	Indo-European	Bubeník 1983	p^{h} t^{h} k^{h} - f θ x	spirantization	X			
Ao (Chungli)	Sino-Tibetan	Gurubaswe-Gowda 1975	p t k - b d g	voicing	X			
Apalai	Carib	Koehn & Koehn 1986	k - g	voicing	X			
Apatani	Sino-Tibetan	Abraham 1985	b - β	spirantization	X			
Arabic (Egyptian)	Afro-Asiatic	Harrell 1957	onset fortis, coda lenis	intercostal pressure	X			
Arabic (Saʕi:di)	Afro-Asiatic	Khalafallah 1969	fortis - lenis	unclear	X			
Arbore	Afro-Asiatic	Harris 1990, Hayward 1984	t' tʃ' k' ɓ̥ ɗ - ʔ	debuccalization				X
Assamese 1	Indo-European	Goswami 1966	ɸ β only _#	distributional property	X			
Assamese 2			p^{h} t^{h} $b^{ɦ}$ - ɸ θ β	spirantization	X			
Babine 1	Na-Dene	Story 1984	s - z	voicing	X			

Babine 2			x - h	debuccalization	X			
Badimaya 1	Australian	Dunn 1988	Stop voicing	voicing	X			
Badimaya 2			Velars rarely voice	distributional property	X			
Badimaya 3			t̪, c - ð, ʒ	spirantization	X			
Balti (Purki)	Sino-Tibetan	Rangan 1975	d̪ - ð; ð ʁ f restricted	spirantization (& dist. Obs)	X			
Bashkir	Altaic	Poppe 1964	b - β	spirantization	X			
Basque (Souletin)	Basque	Hualde 1993	b d g - β ð ɣ	spirantization	X			
Blackfoot	Algic	Frantz 1971	w j - Ø	loss	X			
Bontoc	Austronesian	Reid 1971, Sadenadel 1909	b d g - f ts kʰ	devoicing (& spirantization)	X			
Bulgarian 1	Indo-European	Scatton 1984	vN - ṽ	nasalization & loss	X			
Bulgarian 2			voicing assimilation	voice assimilation				X
Bulgarian 3			final devoicing	devoicing				X
Bulgarian 4			CC - C	degemination	X			
Bulgarian 5			ts x - dz ɣ	voicing	X			
Canela-Krahô 1	Macro-Ge	Popjes & Popjes 1986	l - ɾ	flapping	X			
Canela-Krahô 2			p t k - b d g	voicing	X			
Canela-Krahô 3			j x - h	debuccalization	X			
Car Nicobarese	Austro-Asiatic	Das 1977	r - d	occlusivization	X			
Catalan (Eastern)	Indo-European	Hualde 1992	b d g - β ð ɣ	spirantization	X			
Cuna	Chibchan	Keating et al. 1983	p t k - b d g	voicing	X			
Dahalo	Afro-Asiatic	Tosco 1991	b d - β ð	spirantization	X			
Djabugay 1	Australian	Patz 1991	r - ɾ	flapping	X			
Djabugay 2			p t c k - b d ɟ g	voicing	X			
Efik (Calabar-Creek)	Niger-Congo	Dunstan 1969	b d k - β ɾ ɣ	spirantization	X			

English (American)	Indo-European	Kahn 1976, Chomsky 1964, Joos 1942	t d - ɾ	flapping	X				
English (British)	Indo-European	Milroy et al. 1994	t - ʔ	debuccalization	X				
English (Cardiff) 1	Indo-European	Collins & Mees 1990	b - β	spirantization	X				
English (Cardiff) 2			b d g dʒ - p t k tʃ	devoicing	X				
English (Cockney)	Indo-European	Andrésen 1968	t - ʔ	debuccalization	X				
English (Liverpool)	Indo-European	Wells 1982a	θ̪ ð̪ - t d	occlusivization	X				
English (London, Leeds & Fife)	Indo-European	Harris 1990	t - ʔ	debuccalization	X				
English (New Zealand)	Indo-European	Holmes 1994	t - ɾ	flapping	X				
Estonian (Sourthern)	Uralic	Harms 1962	obstruent voicing	voicing	X				
Faroese	Indo-European	Lockwood 1964	p t c k - b d ɟ g	voicing	X				
Finnish	Uralic	Sulkala et al. 1992	r - ɾ	flapping	X				
French (Périgourdin) 1	Indo-European	Marshall 1984	t - d	voicing	X				
French (Périgourdin) 2			d - ð	spirantization	X				
Gbɛya	Niger-Congo	Smarin 1966	ʔ - Ø	loss	X				
Georgian	South Caucasian	Aronson 1989	v - w	gliding	X				
Gitksan	Penutian	Hoard 1978	p t ts k kʷ q - b d dz g gʷ ɢ	voicing	X				
Gojri	Indo-European	Sharma 1979	CC - C:	shortening	X				
Gooniyandi	Australian	McGregor 1990	p t̪ t ʈ c k - b d̪ d ɖ ɟ g	voicing	X				

Gothic	Indo-European	Bennett 1980	b d g - β ð ɣ	spirantization	X				
Greenlandic (West)	Eskimo-Aleut	Fortescue 1984	q - χ or ʁ	spirantization	X				
Guayabero	Guahiban	Keels 1985	d - θ	spirantization & devoicing	X				
Gujarati 1	Indo-European	Cardona 1965	bʰ dʰ ɖʰ gʰ - Hβ Hð Hɽ Hɣ	spirantization	X				
Gujarati 2			ɖ - ɽ	flapping	X				
Haitian Creole 1	Indo-European	Tinelli 1981	j - Ø	loss	X				
Haitian Creole 2			ʒ - j	gliding	X				
Hebrew (Tiberian) 1	Afro-Asiatic	Malone 1993, Prince 1975	p b t d k g - f v θ ð x ɣ	spirantization	X				
Hebrew (Tiberian) 2			C: - C	degemination	X				
Icelandic	Indo-European	Thrainsson 1978	pp tt kk - ʰp ʰt ʰk	preaspiration	X				
Italian (Southern)	Indo-European	Oftedal 1985	p t k - b̥ d̥ g̊	voicing	X				
Kabardian (Terek)	North Caucasian	Colarusso 1988	ts' - s'	spirantization	X				
Kagate 1	Sino-Tibetan	Hoehlig & Hari 1976	b ~ β	spirantization	X				
Kagate 2			g - lenis	unclear	X				
Kagate 3			r - ɾ	flapping	X				
Kaliai-Kove (Kandoka-Lusi)	Austronesian	Counts 1969	g ~ ɣ	spirantization	X				
Kanakuru	Afro-Asiatic	Newman 1974	p t k - w ɹ ɰ	gliding	X				
Kannada	Dravidian	Chisum 1975, Andronov 1969	k t p - g d b	voicing	X				
Kanuri	Nilo-Saharan	Lukas 1967	b g - β ɣ	spirantization	X				

Karao	Austronesian	Brainard 1994	p t d tʃ dʒ g^w q - ɸ θ l ɾ j w χ	spirantization	X			
Kashmiri	Indo-European	Kachru 1969	ɖ - ɽ	flapping	X			
Kirghiz	Altaic	Hebert & Poppe 1963	β - w	gliding	X			
Korean	Language Isolate	Martin 1992	stops voiced	voicing	X			
Kuna (Paya)	Chibchan	Pike et al. 1986	b d g - β ð ɣ	spirantization	X			
Kupia (Sujanakota)	Indo-European	Christmas & Christmas 1975	p ʈ ɖ - ɸ ƫ ɽ	spirantization	X			
Ladakhi	Sino-Tibetan	Koshal 1976	b d g - β̞ ð ɣ	spirantization	X			
Lahaul (Paʈʈani) 1	Sino-Tibetan	Sharma 1982	p^h - p	deaspiration		X		
Lahaul (Paʈʈani) 2			b d g - b̥ d̥ g̊	devoicing	X			
Lahaul (Paʈʈani) 3			ɖ - ɻ	gliding	X			
Lahaul (Paʈʈani) 4			h weak	deaspiration	X			
Lahaul (Paʈʈani) 5			stops unreleased	unrelease	X			
Lahaul (Paʈʈani) 6			b d g - b^ʔ d^ʔ g^ʔ	glottalization	X			
Lama	Niger-Congo	Ourso & Ulrich 1990	p - w	spirantization/ gliding				X
Lamani	Indo-European	Trail 1970	ɖ - ɽ	flapping	X			
Laotian (Vientiane) 1	Tai-Kadai	Morev et al. 1979	stops unreleased	stop release	X			
Laotian (Vientiane) 2			w j - u i	vocalization	X			
Laotian (Vientiane) 3			C^h, voiced C only in onset	distributional property	X			
Lezgian	North Caucasian	Haspelmath 1993	p k t ts tʃ q - b g d z ʒ ʁ	voicing	X			

Limbu (Phedāppe) 1	Sino-Tibetan	van Driem 1987	p tʃ k - b dʒ g	voicing	X			
Limbu (Phedāppe) 2			p^h t^h k^h - b^h d^h g^h	voicing	X			
Limbu (Phedāppe) 3			t - ʔl (ʔtl)	lateralization	X			
Limbu (Phedāppe) 4			b - w	gliding		X		
Lotha 1	Sino-Tibetan	Acharya 1983	t - d	voicing	X			
Lotha 2			k - g	voicing	X			
Lowland Murut (Poros)	Austronesian	Halle & Clements 1983, Prentice 1971	b d g - β ɾ ɣ	spirantization	X			
Macushi	Carib	Abbott 1991	p t s k - b d z g	voicing	X			
Maidu	Penutian	Shipley 1963	p t k - $p^{ʔ}$ $t^{ʔ}$ $k^{ʔ}$	glottalization	X			
Malayalam 1	Dravidian	Mohanan 1986	p t̪ ʈ k - b d̪ ɖ g	voicing	X			
Malayalam 2			b d̪ ɖ g - β ð̪ ɽ ɣ	spirantization	X			
Manobo (Ata)	Austronesian	Morey & Post 1964	b d g - β ð ɣ	spirantization	X			
Maori	Austronesian	Bauer 1993	k - x/ç	spirantization	X			
Marathi (Halabi)	Indo-European	Singh 1977	ɖ $ɖ^h$ - ɽ $ɽ^h$	flapping	X			
Mataco (Noctenes)	Mataco-Guaicuru	Claesson 1994	ʔ h - Ø	loss	X			
Maxakalí	Macro-Ge	Gudschinsky et al. 1970	stops - vocoids	gliding	X			
Mbabaram	Australian	Dixon 1991	stops voiced	voicing	X			
Middle Chinese	Sino-Tibetan	Pulleyblank 1984	χ - x/h	debuccalization	X			
Middle Korean	Language Isolate	Ramsey 1991	b - β	spirantization	X			
Modern Irish	Indo-European	Ó Siadhail 1989	tense/lax	dialectal variation	X			
Moghamo (Batibo)	Niger-Congo	Stallcup 1978	d g - ɽ ɣ	spirantization	X			

Mohawk (Akwesasne)	Iroquoian	Halle & Clements 1983, Bonvillain 1973	t k - d g	voicing	X			
Mongolian (Kalkha)	Altaic	Poppe 1970	b g (ɢ) - β ɣ (ʁ)	spirantization	X			
Navajo 1	Na-Dene	Kari 1976	x - h	debuccalization	X			
Navajo 2			ɣ j - Ø	loss	X			
Nepali 1	Indo-European	Bandu & Dahal 1971, Acharya 1991	p^h b^h k^h - ɸ β kx	spirantization	X			
Nepali 2			b d - 'lax' b d	laxing	X			
Nepali 3			ɖ $ɖ^ɦ$ - ɽ $ɽ^ɦ$	flapping	X			
Nepali 4			ts^h - h	debuccalization	X			
Nepali 5			h - Ø	loss		X		
Nepali 6			b^h d^h g^h - b d g	deaspiration			X	
Newari	Sino-Tibetan	Nanda 1971	h - Ø	loss	X			
Nez Perce 1	Penutian	Aoki 1970	t k q - s x χ	spirantization				X
Nez Perce 2			h - Ø	loss		X		
Nkore-Kiga	Niger-Congo	Taylor 1985	b - β	spirantization	X			
Northern Corsican 1	Indo-European	Oftedal 1985, Dinnsen & Eckman 1977	p t k - b d g	voicing	X			
Northern Corsican 2			b d g - β ð ɣ	spirantization	X			
Northern Corsican 3			f s - v z	voicing	X			
Northern Corsican 4			v - w	gliding	X			
Numic	Uto-Astecan	Manaster Ramer 1993	*kk > k	degemination	X			
Nyawaygi	Australian	Dixon 1983	r - d	occlusivization	X			
Páez	Paezan	Gerdel 1985	x - h	debuccalization	X			

Panyjima (Mijaranypa) 1	Australian	Dench 1991	ʈ - ɽ	flapping	X			
Panyjima (Mijaranypa) 2			r - ɾ or ɹ	flapping/ gliding	X			
Panyjima (Mijaranypa) 3			no r /_#	distributional property	X			
Pawnee 1	Caddoan	Parks 1976	h - Ø	loss				X
Pawnee 2			ɾ - Ø	loss				X
Pipil (Cuisnahuat)	Uto-Aztecan	Campbell 1985	w - h	debuccalization				X
Proto-Ainu	Language Isolate	Vovin 1993	*g > h	debuccalization	X			
Proto-Bantu	Niger-Congo	Halle & Clements 1983, Greenberg 1948	b d g - β l ɣ	spirantization	X			
Proto-Germanic 1	Indo-European	Meillet 1970, Prokosch 1933	b^h d^h g^h > β ð ɣ	spirantization	X			
Proto-Germanic 2			p^h t^h k^h > f θ x	spirantization	X			
Ptolemaic Greek	Indo-European	Teodorsson 1977	g > ɣ	spirantization	X			
Punjabi 1	Indo-European	Gill & Gleason Jr 1969	geminates only after central V	distributional property	X			
Punjabi 2			p^h - ɸ	spirantization	X			
Purki 1	Sino-Tibetan	Rangan 1979	d̪ - ð	spirantization	X			
Purki 2			d - ɽ	flapping	X			
Purki 3			assymetry in C inventory	distributional property	X			
Quechua	Quechuan	Whitley 1978, Bills et al. 1969	k q - x χ	spirantization	X			
Russian 1	Indo-European	Wade 1992	voicing assimilation	voice assimilation				X

Russian 2			degemination	degemination	X			
Saek	Tai-Kadai	Hudak 1993	ɣ ~ g	occlusivization	X			
Sanuma 1	Yanomam	Borgman 1990	p t ts k - b d dz g	voicing	X			
Sanuma 2			d dz - ɾ z	spirantization	X			
Sanuma 3			h - Ø	loss				X
Sawai	Austronesian	Whisler 1992	d - ɾ	flapping	X			
Sekani	Na-Dene	Hargus 1985	s ɬ ç x ʍ - z l ʝ ɣ w	voicing	X			
Senoufo 1	Niger-Congo	Mills 1984	p t k - b d g	voicing	X			
Senoufo 2			b d g - β ɾ ɣ	spirantization	X			
Serbo-Croatian	Indo-European	Partridge 1988	voicing assimilation	voice assimilation				X
Shapsug (Düzce)	North Caucasian	Smeets 1984	articulatory force varies	unrelease	X			
Shina 1	Indo-European	Rajapurohit 1983	b d g - ƀ đ ǥ	spirantization	X			
Shina 2			ɖ - ɽ	spirantization	X			
Shina 3			r - ɾ	flapping	X			
Shoshone (Tümpisa) 1	Uto-Aztecan	Dayley 1989	p t c k kʷ - ɸ θ ç x xʷ	spirantization	X			
Shoshone (Tümpisa) 2			p t c k kʷ - β ð ʝ ɣ ɣʷ	spirantization & voicing	X			
Shoshone (Tümpisa) 3			t - ɾ	flapping	X			
Shoshone (Tümpisa) 4			p t c k kʷ - b d ɟ g gʷ	voicing	X			
Shoshone (Tümpisa) 5			m n - w̃ j̃	gliding	X			
Shoshone (Tümpisa) 6			ʔ h - Ø	loss	X			
Somali	Afro-Asiatic	Armstrong 1964	b d ɖ g - β ð ɽ ɣ	spirantization	X			
Sotho (Southern) 1	Niger-Congo	Doke & Mofokeng 1957	l - d	delateralization	X			
Sotho (Southern) 2			b > β	spirantization	X			

Southern Tati (Chali) 1	Indo-European	Yar-Shater 1969	q - ʁ	spirantization	X				
Southern Tati (Chali) 2			h - Ø	loss	X				
Southern Tati (Eshtehardi)	Indo-European	Yar-Shater 1969	v - w	gliding	X				
Southern Tati (Takestani)	Indo-European	Yar-Shater 1969	g - ɣ	spirantization	X				
Southern Tati (Xiaraji)	Indo-European	Yar-Shater 1969	v - w	gliding	X				
Spanish (Andalusian)	Indo-European	Romero-Gallego 1995	b d g - β̞ ð̞ ɣ	spirantization	X				
Spanish (Mexico City) 1	Indo-European	Harris 1969	b d g - β ð ɣ	spirantization	X				
Spanish (Mexico City) 2			r - ɾ	tapping					X
Tahltan (Iskut)	Na-Dene	Nater 1989	C̥ - C	voicing	X				
Taiwanese	Sino-Tibetan	Hsu 1995	p t k - β ɾ ɣ	spirantization	X				
Tamazight Berber (Ayt Ayache) 1	Afro-Asiatic	Abdel-Massih 1971	b - β	spirantization	X				
Tamazight Berber (Ayt Ayache) 2			k g > x ɣ	spirantization	X				
Tamazight Berber (Ayt Ayache) 3			spirantization blocked in C:	distributional property	X				
Tamazight Berber (Ayt Ndhir) 1	Afro-Asiatic	Saib 1976	bb tt dd - β θ ð	spirantization	X				
Tamazight Berber (Ayt Ndhir) 2			ðˤ zˤ - tˤtˤ sˤsˤ	voicing	X				
Tamazight Berber (Ayt Seghrouchen)	Afro-Asiatic	Abdel-Massih 1971	b - β	spirantization	X				

Tamil	Dravidian	Annamalai 1975	p t ts ʈ k - β ð dz ɽ ɣ	spirantization	X			
Tatar	Altaic	Poppe 1968	b d g - β ð ɣ	spirantization	X			
Tauya 1	Trans-New Guinea	MacDonald 1990	k k^{w} > ʔ $ʔ^{w}$	debuccalization	X			
Tauya 2			r - t	occlusivization	X			
Thai	Tai-Kadai	Noss 1964	m n ŋ w j shorten	shortening	X			
Tharu (Chitwan) 1	Indo-European	Leal 1972	h - Ø	loss			X	
Tharu (Chitwan) 2			p^{h} b ɖ $ɖ^{h}$ dʒ - ɸ β ɾ $ɾ^{h}$ ʒ	spirantization	X			
Tigrinya 1	Afro-Asiatic	Kenstowicz 1982	k q - x χ	spirantization	X			
Tigrinya 2	Afro-Asiatic	Kenstowicz 1982	Spirantization blocked in C:	distributional property	X			
Toba Batak	Austronesian	Hayes 1986a,b	p t k - ʔ	debuccalization				X
Tojolabal 1	Mayan	Furbee-Losee 1976	r - ɣ	spirantization	X			
Tojolabal 2			g - ɣ	spirantization	X			
Tojolabal 3			h w j - Ø	loss				X
Totonac (Misantla) 1	Totonacan	MacKay 1994	q - χ	spirantization	X			
Totonac (Misantla) 2			p t ts tʃ k q - b d dz dʒ g ɢ	voicing	X			
Tsou (Tfuya)	Austronesian	Wright 1996	^{ʔ}ɗ - ^{ʔ}l	flapping & lateralization	X			
Turkana	Nilo-Saharan	Dimmendaal 1983	t - s	spirantization				X
Turkish 1	Altaic	Underhill 1976	g - Ø, j, ɣ	loss (or gliding/ spirantization)	X			
Turkish 2			v - β, w	fronting & gliding	X			
Tzeltal	Mayan	Kaufman 1971	b d g - β ð ɣ	spirantization	X			
Uradhi	Australian	Crowley 1983	*p *t *k > β ð ɣ	spirantization	X			

Urdu	Indo-European	Beg 1988, Bokhari 1985	ɖ ɖʰ ~ ɽ ɽʰ	flapping	X			
Urubu-Kaapor	Tupi	Kakumasu 1986	p t k - b d g (ɣ)	voicing	X			
Uyghur	Altaic	Hahn 1991	b k g q ɢ - ɸ/β xʲ ɣ χ ʁ	spirantization	X			
Uzbek	Altaic	Sjoberg 1963	b p q ~ w ɸ x/ɣ	spirantization				X
Vietnamese (Vinh)	Austro-Asiatic	Emeneau 1951	ɣ - g	occlusivization	X			
Warndarang 1	Australian	Heath 1980	p t ʈ c k - b d ɖ ɟ g	voicing	X			
Warndarang 2			p c - w j	gliding	X			
West Tarangan 1	Austronesian	Nivens 1992	j w - g dʒ	occlusivization	X			
West Tarangan 2			k - ʔ	debuccalization	X			
Wiyot	Algic	Teeter 1964	no Cʰ / _#	distributional property	X			
Yakut 1	Altaic	Krueger 1962, Barashkov 1953	p - b	voicing	X			
Yakut 2			k - χ	spirantization	X			
Yakut 3			no vd geminate obst	distributional property	X			
Yana 1	Hokan	Sapir & Swadesh 1960	p t ts k partially voice	voicing	X			
Yana 2			b - w	gliding		X		
Yindjibarndi 1	Australian	Wordick 1982, O'Grady 1966	p t̪ ʈ k - w j j w	gliding	X			
Yindjibarndi 2			ʈ - ɽ	flapping	X			
Yolngu (Djapu)	Australian	Murphy 1983	t̪ c - j; k p - w	gliding				X
Yuman	Hokan	Wares 1968	c > s	spirantization	X			

APPENDIX B

Glossary

ALLOPHONE: a phonetic variation of a PHONEME. In literature an allophone is usually indicated using square brackets: [allophone].

DISTRIBUTIONAL PROPERTY: a statement about the (usually limited) context in which certain sounds can appear in a language.

FORTITION (or STRENGTHENING): the strengthening of a sound in a particular environment, requiring a somewhat greater effort of articulation. The theoretic opposite of LENITION.

FUNCTION WORDS: words that serve a grammatical purpose in a language. These are the more common words in a language that are usually contextually predictable.

FUNCTIONAL CONSIDERATIONS: considerations of meaning and how it is affected by sound substitutions, which may either obliterate meaning distinction (i.e., lead to PHONOLOGICAL NEUTRALIZATION) or maintain it. The primary *function* of language is to convey meaning.

FUNCTIONAL LOAD: the number of elements in the lexicon that depend on a particular opposition in order to be distinguished from others.

HOMOPHONY: total obliteration of distinction between words (obliteration of LEXICAL DISTINCTION). E.g., in many dialects of Modern Hebrew ʕ → ʔ; MINIMAL PAIRS such as ʔim 'if' and ʕim 'with' end up as homonyms because ʕim → ʔim, and the distinction between 'if' and 'with' is obliterated.

IMPLICATIONAL HIERARCHY: if a language has voiced stops it has voiceless ones (Westbury and Keating 1986); thus an inventory that includes voiced but not voiceless stops is considered unnatural (see discussion of Badimaya, §3.11).

LENITION (or WEAKENING): a cover term for a variety of sound changes driven by phonetic factors, especially "ease of production" of sounds in certain environments.

LEXICAL DISTINCTION: distinction between lexical items (words). The obliteration of this distinction leads to HOMOPHONY.

LOSS (or DELETION): where a sound is completely dropped (i.e., realized as Ø) in a certain environment.

MARKEDNESS: a mostly generative theory-internal measure of "naturalness" of a linguistic element, where the less marked the more "natural" an element is.

MINIMAL PAIRS: a pair of words that are distinguished only by a difference in one PHONEME (e.g., Hebrew ʔani 'I' vs. ʕani 'poor'). Such pairs are proof that a sound in a language is PHONEMIC (i.e., represents meaning) because it changes the meaning of a word.

PHONEME: a sound which serves a contrastive function in a language. In literature a phoneme is usually indicated using virgules: /phoneme/.

PHONEMIC OVERLAP: an intersection of PHONEMES where one PHONEME is substituted with a sound that represents another PHONEME, but NEUTRALIZATION is avoided because the second PHONEME is substituted with a third sound in the same environment.

PHONETIC NEUTRALIZATION: the obliteration of distinction between two sounds so that they are pronounced, and sound, the same.

PHONETICALLY CONDITIONED SOUND CHANGES: changes affected by physical properties such as the shape and movement of the vocal organs.

PHONOLOGICAL NEUTRALIZATION: PHONETIC NEUTRALIZATION that leads to confusion between PHONEMES because it may not be possible to determine which PHONEME is represented by a sound in a given context in some language. This is the only type of neutralization that can result in HOMOPHONY, and therefore the only one with potentially crucial consequences for communication.

VOICING ASSIMILATION: a phenomenon where sounds are either voiced, or devoiced depending on the context, to match the voicing feature of another sound in their vicinity.

Index

For Product Safety Concerns and Information please contact our EU
representative GPSR@taylorandfrancis.com
Taylor & Francis Verlag GmbH, Kaufingerstraße 24, 80331 München, Germany

www.ingramcontent.com/pod-product-compliance
Ingram Content Group UK Ltd.
Pitfield, Milton Keynes, MK11 3LW, UK
UKHW022116190926
475702UK00013B/1466

* 9 7 8 0 4 1 5 8 6 5 1 4 2 *